PEARLS OF WISDOM SERIES

Medical Biochemistry

PEARLS OF WISDOM

Duane C. Eichler, PhD

Professor, Biochemistry and Molecular Biology
University of South Florida
College of Medicine

JONES AND BARTLETT PUBLISHERS
Sudbury, Massachusetts
BOSTON   TORONTO   LONDON   SINGAPORE

Jones and Bartlett Publishers

World Headquarters
Jones and Bartlett Publishers
40 Tall Pine Drive
Sudbury, MA 01776
978-443-5000
info@jbpub.com
www.EMSzone.com

Jones and Bartlett Publishers Canada
6339 Ormindale Way
Mississauga, ON L5V 1J2
Canada

Jones and Bartlett Publishers
International
Barb House, Barb Mews
London W6 7PA
United Kingdom

Jones and Bartlett's books and products are available through most bookstores and online booksellers. To contact Jones and Bartlett Publishers directly, call 800-832-0034, fax 978-443-8000, or visit our website, www.jbpub.com.

Substantial discounts on bulk quantities of Jones and Bartlett's publications are available to corporations, professional associations, and other qualified organizations. For details and specific discount information, contact the special sales department at Jones and Bartlett via the above contact information or send an email to specialsales@jbpub.com.

Production Credits
Chief Executive Officer: Clayton E. Jones
Chief Operating Officer: Donald W. Jones, Jr.
President, Higher Education and Professional Publishing: Robert W. Holland, Jr.
V.P., Sales and Marketing: William J. Kane
V.P., Production and Design: Anne Spencer
V.P., Manufacturing and Inventory Control: Therese Connell
Acquistions Editor, Science: Cathleen Sether
Managing Editor, Science: Dean DeChambeau
Editorial Assistant, Science: Molly Steinbach
Production Editor: Karen Ferreira
Marketing Manager: Andrea DeFronzo
Composition: N.K. Graphics
Illustration: N.K. Graphics
Text and Cover Design: Anne Spencer
Printing and Binding: Courier Stoughton
Cover Printing: Lehigh Press

ISBN: 0-7637-3525-6
ISBN-13: 978-0-7637-3525-8

Library of Congress Cataloging-in-Publication Data

Eichler, Duane C.
    Medical biochemistry : pearls of wisdom / Duane C. Eichler.
        p. ; cm.
    ISBN 0-7637-3525-6
    1. Biochemistry. 2. Metabolism.   I. Title.
    [DNLM: 1. Biochemistry. 2. Metabolism.   QU 4 E338m 2006]
QP514.2.E35 2006
612.3'9—dc22
6048
                            2005021202

Printed in the United States of America
10 09 08 07 06   10 9 8 7 6 5 4 3 2 1

Contents

Introduction

Medical Biochemistry Pearls of Wisdom is designed to help you prepare for your course, board, and recertification exams. *Pearls'* unique format differs from all other review and test preparation texts. What follows is a brief introduction to its purpose, format, limitations, and intended use.

The primary intent of *Pearls* is to serve as a study aid to improve performance on medical biochemistry examinations. With this goal in mind, the text is written in rapid-fire, question and answer format. You will receive immediate gratification with a correct answer. Misleading or confusing multiple-choice "foils" are not provided, thereby eliminating the risk of assimilating erroneous information that made an impression. Another advantage of this format is that you will either know or not know the answer to a given question. This results in active learning, rather than the passive review of studying multiple choice questions.

Questions themselves often contain a pearl reinforced in association with the question and answer. Additional information not requested in the question may be included in the answer. The same information is often sought in several different questions. Emphasis has been placed on evoking both trivia and key facts that are easily overlooked, are quickly forgotten, and yet somehow always seem to appear on exams.

It may happen that upon reading an answer you may think: "Why is that?" or, "Are you sure?" If this happens to you, go check! Truly assimilating these disparate facts into a framework of knowledge absolutely requires further reading in the surrounding concepts. Information learned as a response to seeking an answer to a particular question is much better retained than information that is passively read. Take advantage of this. Use *Pearls* with your preferred source texts nearby and open, or, if you are reviewing without your texts handy, mark questions for further investigation.

Pearls has limitations. There may be conflicts among texts on medical biochemistry. By its very nature, soon after publication many of the concepts will not represent the cutting edge of biochemistry. With these limitations in mind, *Pearls* risks accuracy by aggressively pruning complex concepts down to the simplest kernel. New research and practice occasionally deviates from that which likely represents the "right" answer for test purposes. In such cases we have selected the information that we believe is most likely "correct" for test purposes. This text is designed to maximize your score on a test. Refer to your most current sources of information, your mentors, your protocols and your instructor for direction on current practice.

Pearls is designed to be used, not just read. It is an interactive text. Use a 3 by 5 card and cover the answers; attempt all questions. A study method we strongly recommend is oral, group study, preferably over an extended meal. The mechanics of this method are simple and no one ever appears stupid. One person holds *Pearls*, with answers covered, and reads the question. Each person, including the reader, says "Check!" when he or she has an answer in mind. After everyone has "checked" in, someone states his or her answer. If this answer is correct, on to the next one. If not, another person states his or her answer, or the answer can be read aloud. Usually, the person who checks in first gets the first shot at stating the answer. Try it—it's almost fun!

Pearls is also designed to be re-used several times to allow, dare we use the word, memorization. If you are a pessimist, we suggest putting a check mark next to a question every time it is missed. If you are an optimist, place a check mark when the question is answered correctly once; skip all questions with check marks thereafter. Utilize whatever scheme you prefer.

The publisher and I welcome your comments, suggestions, and criticism. Great efforts have been made to verify these questions and answers. There will be answers we have provided that are at variance with the answer you would prefer. This is most often the result of differences between the original source and the source you have chosen to use. Please make us aware of any errata you find. We hope to make continuous improvements in future editions and would greatly appreciate any input with regard to format, organization, content, presentation, or about specific questions.

Study hard and good luck!

D.E.

DEDICATION

To my wife, Sandy Marie, whose love, support, and encouragement allowed me to advance confidently in the direction of my dreams; and to my father, Arthur O. Eichler, who instilled in me a need to ask questions and an enthusiasm for learning.

Cell Structure and Function

Compartments and Function

What is the essential structural difference that distinguishes eukaryotic from prokaryotic cells?

In eukaryotic cells, membranes partition the cell into functionally distinct compartments. Compartmentation of function allows for the increased specialization and complexity of multicellular eukaryotic organisms.

What advantages does the partitioning by compartments offer to eukaryotic cells?

Compartments offer at least two obvious and important functions. The first is to maximize cellular efficiency by allowing different chemical reactions that require different environments (pH, ionic strength, etc.) to occur simultaneously in the cell, and to allow metabolic pathways involved in the synthesis and degradation of the same compound to be physically partitioned. Second, compartments can be used to enhance regulation. For example, the transfer of a metabolite from one compartment to another can be used as a key step in controlling the rate of that process.

What are the general features of cellular structure?

From the point of a simple overview, seven major compartments are common to most eukaryotic cells (Figure 1.1).

1. Cytosol
2. Mitochondria
3. Rough endoplasmic reticulum cisternae
4. Smooth endoplasmic reticulum cisternae
5. Lysosomes
6. Peroxisomes
7. Nucleus

Throughout any discussion of medical biochemistry, the relationship of a metabolic process to the organizational structure within the cell is important to recognize, and it will become apparent that metabolic patterns of eukaryotic cells are

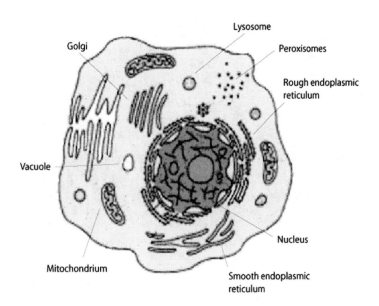

Figure 1.1 General features of cell structure. A schematic view represents the major structural compartments of a eukaryotic cell.

markedly affected by the presence of compartments. Glycolysis, the pentose phosphate shunt pathway, and fatty acid synthesis take place in the cytosol, whereas fatty acid oxidation, the citric acid cycle, and oxidative phosphorylation are carried out in mitochondria. Some processes such as gluconeogenesis, urea synthesis, and even pyrimidine biosynthesis depend on the interplay of reactions that occur in two compartments: the cytosol and mitochondria.

What role does the cytoplasmic membrane play other than to act to partition the biological constituents of the cell from the immediate environment?

In the broadest sense, the cell surface occupies a key position in the economy of the cell as a mediator between the cell and the environment. Specifically, the cytoplasmic membrane acts as a shape constraint, as a selective permeable barrier (transport or at the anatomical scale, membranes and their selective function play a major role in the uptake of food material from the gut, the selective removal of wastes from the blood in the kidney, the blood–brain barrier, neuron transmission, etc.), as a site for receptors (hormones, etc.), and in important cell–cell interactions (contact inhibition, cell adhesion, etc.).

Why is the cytoplasmic membrane structure called a lipid bilayer?

This is due to the defined arrangement of phospholipids into a bilayer, with the polar phosphorous portion of the phospholipid on the outside and the apolar fat portion orienting inward. Within the apolar region, cholesterol molecules are found, whereas proteins, depending on their own bias, may distribute in a variety of places within this matrix. There are also sugar components, carbohydrates, that are primarily found on the outer surface attached to proteins or complex fats (Figure 1.2).

How do proteins that are associated with the cytoplasmic membrane differ?

Proteins that are major molecular constituents of membranes can be divided into two groups. Integral proteins are directly incorporated within the lipid bilayer, whereas peripheral proteins exhibit a looser association with mem-

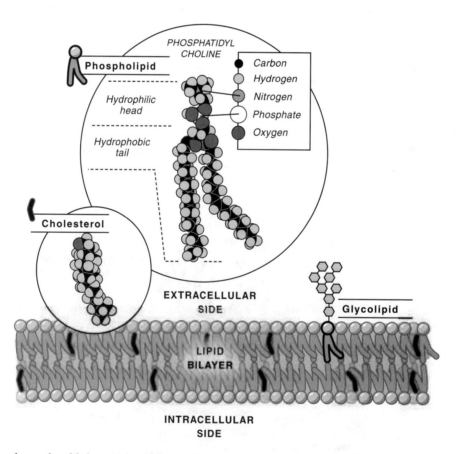

Figure 1.2   Plasma membrane lipid bilayer. The diagram illustrates the basic structural characteristics of the membrane lipid bilayer. The amphipathic phospholipids contain a hydrophilic head and hydrophobic tails. This membrane structure is quite stable because of the hydrophobic interaction of the hydrocarbon chains and attraction of polar groups to water on the outer surface.

brane surfaces. The loosely bound peripheral proteins can be easily extracted from cell membranes with salt solutions, whereas integral proteins can be extracted only by drastic methods that use detergents. Some integral proteins span the membrane one or more times from one side to the other. The extracellular portion of integral and peripheral membrane proteins is generally glycosylated. The intracellular portion of membrane proteins is bound to cytoskeletal components (Figure 1.3).

Do regions of the membrane differ in composition and structure?

There are regions of the cytoplasmic membrane enriched in cholesterol and sphingolipids. These are known as "lipid rafts," and these sites are responsible for cellular functions such as vesicular trafficking and signal transduction.

How do lipid rafts function?

A lipid raft is a precursor to a caveola, which is a structure that is predominant in tissue such as fibroblasts, adipocytes, endothelial cells, and muscle. A protein called caveolin binds to cholesterol in the lipid raft. There are at least three types of caveolins numbered 1, 2, and 3. Caveolae tend to concentrate signaling molecules such as Src-like tyrosine kinases, G protein, and nitric oxide synthase.

What are some of the functional properties of intracellular membranes?

An intracellular membrane system called the endoplasmic reticulum acts to partition the cytoplasm effectively into two phases separated by a single membrane surface. This relatively large area of membrane surface can be accommodated within a limited cytoplasmic volume. Surface-limited reactions can therefore be carried out with increased efficiency. Optimal concentrations and spatial interrelationships can be effectively maintained for enzyme systems attached to these intracellular membranes. This membrane system also provides a complex of interconnecting channels (cisternae) throughout the cell that can be used for the transport of material and a mechanism for separating newly synthesized molecules that belong in the cytosol from those that do not. Thus, the endoplasmic reticulum provides an important internal surface for enzyme systems as well as providing an internal channeling system to vector agents for the processing and transport of a defined class of proteins.

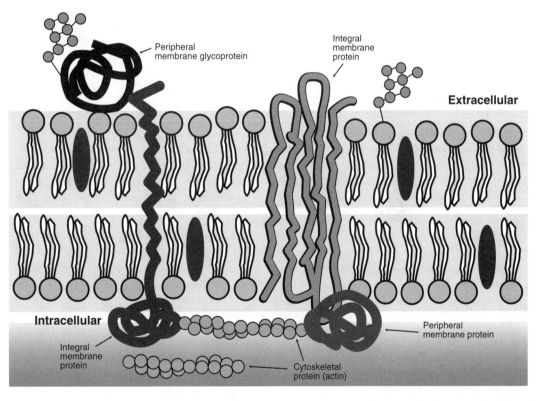

Figure 1.3 Plasma membrane components. The schematic illustrates the multiple types of proteins which bind or interact with the membrane lipid bilayer. Single or multiple transmembrane segments demonstrate integral membrane protein interactions, whereas peripheral membrane proteins bind to the membrane surface and/or integral membrane protein.

Is there a significant functional difference between the smooth and rough endoplasmic reticulum other than appearance?

The rough endoplasmic reticulum refers to the granular appearance of this membrane structure. This granular surface is due to numerous ribosomes associated with the membrane surface. These ribosomes are actively translating mRNA in the process of protein synthesis. The protein being synthesized is destined for export or to sorting to specific regions or compartments of the cell. The smooth endoplasmic reticulum is distinguished simply by the lack of ribosomes. Various important metabolic enzymes are associated with the smooth endoplasmic reticulum such as a group of enzymes that make up an electron transport system completely independent of the mitochondrion, and a key regulatory enzyme for cholesterol biosynthesis is also associated with the smooth endoplasmic reticulum.

What cellular structure plays a key role in the sorting and routing of cellular materials?

The Golgi apparatus or Golgi complex is the separation site of intracellular transport routes involved in the export of proteins from the endoplasmic reticulum. Here, components of the plasma membrane, secretory granules, and lysosomes are sorted and packaged into separate types of transport vesicles for delivery to their appropriate cellular destination. The Golgi stack has functional, as well as topologic, polarity. Vesicles with proteins enter the stack at its *cis* (entry) face and, at least in the case of plasma membrane and secretory proteins, depart at the opposite *trans* (exit) face. Thus, the cisternae at the *cis* and *trans* ends are biochemically distinct, differing in the kinds of proteins responsible for the entry and exit processes.

What organelle plays a key role in energy metabolism in the cell?

The essential functions of mitochondria are devoted to the energy needs of the cell. The respiratory assembly is an integral part of the inner mitochondrial membrane and cristae. Reactions of the TCA (Tricarboxylic acid) cycle and its interrelationship in energy production and oxidative phosphorylation are associated with the respiratory assembly. Specific protein carriers transport molecules such as ADP (Adenosine diphosphate) and long-chain fatty acids across the inner mitochondrial membrane.

What are the specific structural features of the mitochondrion, and what role do they play in the function of this organelle?

Cristae are produced by the folding of the inner mitochondrial membrane into a series of internal ridges. The respiratory assembly is an integral part of the inner mitochondrial membrane and cristae. Reactions of the TCA cycle and the oxidation of fatty acids occur in the matrix, which is the internal volume of the mitochondrion. The inner membrane is intrinsically impermeable to nearly all ions and most uncharged molecules. Therefore, specific protein carriers transport molecules such as ADP and long-chain fatty acids across the inner mitochondrial membrane. The outer membrane is quite permeable to most small molecules and ions.

Why is the mitochondrion considered, in a sense, to be "semiautonomous?"

Mitochondria contain their own DNA, RNA, and protein translational machinery. Thus, this organelle has the capacity to define and maintain some, but not all, of its own functions. Diseases that result from inherited defects in mitochondrial genes are therefore inherited from the mother, because the egg is the source of mitochondria.

What cellular organelle plays a key role in the turnover of most cellular components?

Lysosomes contain many types of degradative enzymes that are specialized for the orderly destruction of cellular components. It is important to consider that most cellular components turn over. In other words, biological compounds are constantly being synthesized and degraded. This process is finely regulated so that the proper "steady-state" level of each component is maintained, and lysosomes are an integral part of this process.

What class of diseases result from defects in a particular degradative lysosomal enzyme?

The importance of lysosomal function is best exemplified by a class of diseases known as "lysosomal storage diseases," which arise from defects in a particular degradative lysosomal enzyme. A genetic defect in a degradative enzyme results in an accumulation of substrate inside the lysosome. This results in cellular dysfunction and eventually in cell death. More than 48 lysosomal hydrolase or lysosomal membrane transport deficiencies have been described. Almost all are autosomal recessive in inheritance (Figure 1.4).

What is Tay-Sachs disease?

Tay-Sachs disease is one of a group of heterogeneous lysosomal storage diseases, the G_{M2} gangliosidoses, that results from the inability to degrade a sphingolipid, G_{M2} ganglioside. The biochemical lesion is a marked deficiency of hexosaminidase A. Although the enzyme is ubiquitous, the disease has its clinical impact almost solely on the brain, the predominant site of G_{M2} synthesis.

What organelle serves an essential purpose of metabolizing hydrogen peroxide?

Most eukaryotic cells of mammalian origin have a defined cellular organelle, peroxisomes (microbodies), that contain several enzymes that either produce or use hydrogen peroxide. The organelles are usually smaller than lysosomes and are spherical or oval in shape, with a granular matrix and in some cases a crystalline inclusion termed a *nucleoid*. Peroxisomes are the most numerous in liver and kidney cells. The matrix of all microbodies contains catalase, a heme protein that combines two molecules of H_2O_2 (hydrogen peroxide) to produce O_2 and H_2O, and peroxidases, which are also heme enzymes that combine H_2O_2 with a hydrogen donor substrate resulting in the oxidation of the donor and the formation of H_2O.

$$\textit{Catalase}$$
$$2H_2O_2 \rightarrow 2H_2O + O_2$$

$$\textit{Peroxidase}$$
$$RH_2 + H_2O_2 \rightarrow R + 2H_2O$$

What disease results from a deficiency in the biogenesis of peroxisomes?

Although the peroxisomal membrane forms in Zellweger syndrome, there is a deficiency in the receptor that targets enzymes to the peroxisomes. As a result, peroxisomal function is compromised; thus, Zellweger syndrome presents with serious neurologic deficits during infancy, and afflicted children usually die during their first or second year. Because one role of the peroxisome involves the enzymatic machinery necessary for the degradation of very long-chain fatty acids via β-oxidation, accumulation of the very long-chain fatty acids (> C:20) is observed in these infants.

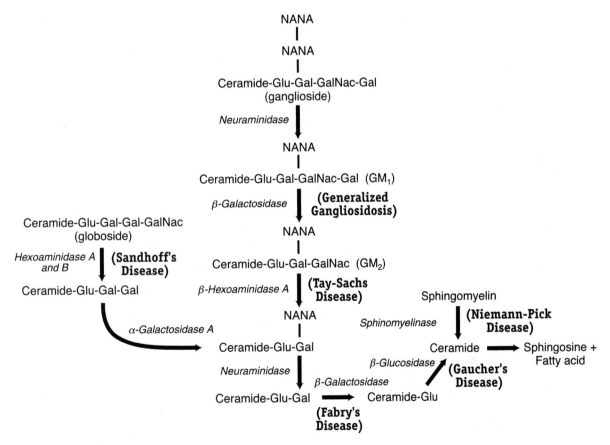

Figure 1.4 The ordered lysosomal catabolism of sphingolipids. Typically lysosomes of histocytes or macrophages of the reticuloendothelial system, located primarily in liver, spleen, and bone marrow, are responsible for the orderly degradation of sphingolipids. Sphingolipidoses caused by a genetically determined enzyme deficiency are indicated by parenthesis and bold print.

What other internal structures of the cell affect cellular function?

The cell contains a microtrabecular system that is essential for the coordinated activities of different parts of the cell. In this role, the cytoskeleton is a major factor in providing the nonhomogeneity of the cytoplasm that distinguishes a cell from an aqueous solution.

What are the basic components of the cytoskeleton?

The cytoskeleton consists of microtubules, intermediate filaments, and microfilaments. Biochemical studies, involving the extraction of cytoskeletal proteins from cells with detergents and salts, showed that each class of filaments has a unique protein organization. The basic component of microtubules is tubulin; intermediate filaments contain a number of substances, including keratin, desmin, and vimentin, whereas microfilaments are essentially composed of actin. The functions of these structures are greatly influenced by a number of other associated molecules.

What are the general functions of microfilaments in nonmuscle cells?

They form cross-linked bundles that provide mechanical support for various cellular structures and extensions. Together with myosin, microfilaments form the diverse contractile systems thought to be responsible for many cellular movements. Examples are microvilli, bundles of actin filaments that cover the exposed surfaces of many kinds of epithelial cells where cellular function requires a maximum surface for adsorption. Microspikes are composed of actin filaments and act as sensory devices or feelers by which cells explore their environment. The contractile ring consists of actin bundles with myosin, capable of contraction in cell division. Stress fibers are bundles of actin filaments that lie close to the cell surface and terminate at specialized regions of the plasma membrane known as adhesion plaques.

How do microtubules affect the shape and movement of cells?

Microtubules consist of a polymer of the protein tubulin that can be rapidly assembled and disassembled depending on the needs of the cell and are important in the formation of the mitotic apparatus spindle fibers, flagella, cilia, etc. As a result of rapid polymerization and depolymerization, thrusting out and retracting rigid tubular structures change the shape of a cell and are important in movement, aligning internal structures, and in producing localized changes in the surface of the cell.

What are intermediate filaments?

Intermediate filaments are tough, durable protein fibers that appear as straight or gently curving arrays in electron micrographs. They seem to be particularly prominent in those cells that have parts subject to stress, corresponding to their major function, which is to provide mechanical support for the cell. The types and composition of intermediate filaments vary between cell types, species, etc. The structure of intermediate filaments does not fluctuate between assembly and disassembly states such as microtubules and microfilaments. Unlike actin and tubulin, the assembly and disassembly of intermediate filament monomers are regulated by phosphorylation.

How do intermediate filaments differ from microfilaments?

Intermediate filaments consist of fibrous polypeptides that vary greatly in size (40,000 to 200,000 Da). They are generally defined by cellular extraction procedures and are the insoluble fibers left after high and low salt extraction with ionic detergents. Their structure and assembly are similar to collagen.

What structure is typically the most prominent organelle in eukaryotic cells?

The nucleus is the most prominent organelle in a wide variety of eukaryotic cells. The Greek word *karyon* means nucleus.

What essential role does the nucleus serve in affecting cellular function?

The nucleus has a vital role in directing protein synthesis through which it dominates and controls the structure and function of the cell.

What two membrane systems define the boundaries of the nuclear organelle?

The nuclear envelope partitions the nucleus from the cytoplasm and is composed of two separate membranes, each of which shows the characteristic trilaminar membrane structure. The inner nuclear membrane forms the limit of the nuclear contents and is separated by a space of 500 Å from the outer nuclear membrane. The width of this gap varies in different cells and is called the perinuclear space (cisternae).

How do large molecules enter and leave the nucleus?

Nuclear pores are involved in macromolecular transport between the nucleus and cytoplasm. A diaphragm with a dense collar (annulus) is part of the nuclear pore structure.

What are the general components of the nucleus?

The nucleus contains the bulk of cellular DNA, appreciable quantities of RNA, basic and acidic proteins of which a group of small basic proteins, called histones, are the most predominant species.

What relationship does the nucleolus have to nuclear function?

The nucleolus represents the aggregation of genes and specialized machinery involved in ribosomal RNA synthesis and maturation.

How does the structural organization of DNA in the nucleus relate to its ability to control cell function?

The nuclear organelle probably represents the highest form of structural organizational requirements. A human cell contains sufficient DNA to stretch 3 meters fully extended and is able to accommodate this DNA within roughly a 5-micron diameter space of the nucleus. Besides the enormous task of packaging this DNA (chromatin), the nucleus possesses the machinery to permit specific expression of regions (transcription) of this DNA for normal cell function. In addition, the nucleus contains the machinery for the maintenance (DNA repair) of the integrity of the genome, which is paramount to the survival and normal function of the cell. Finally, for those cells that are in active growth, machinery for replication of DNA and the mitotic process is also an integral part of the nucleus.

Chromatin Structure

What is chromatin?

Chromatin refers specifically to the complex of DNA and protein found in nuclei of eukaryotic cells.

What are the general features of chromatin in the nucleus of nondividing cells (interphase)?

Two forms of chromatin material can be identified in the interphase nucleus. Chromosomal material that returns to a dispersed condition of interphase chromatin is referred to as euchromatin. Chromatin material that remains in its condensed state is referred to as heterochromatin.

How does the packing of chromatin as either heterochromatin or euchromatin relate to transcriptional activity (gene expression)?

Heterochromatin is essentially transcriptionally inert and is not used to direct protein synthesis. On the other hand, euchromatin has regions that are active in RNA synthesis. This difference supports the role that the packaging of DNA can affect gene transcription. Any chromosome may have both heterochromatic and euchromatic regions. The centromeres and telomeres of chromosomes represent heterochromatic regions of chromosomes.

Is heterochromatin all the same?

Actually, two classes or types of heterochromatin appear to share the same characteristic condensed morphology and lack of transcriptional activity. Constitutive heterochromatin represents regions of densely staining material that is found essentially at the same position in both members of a homologous chromosome pair. Therefore, the location of constitutive heterochromatin is characteristic within a chromosome set. Facultative heterochromatin in mammalian tissue represents the inactivation of one of the X chromosomes in females by condensation into heterochromatin. Such condensed chromosomes are known as sex chromosome bodies, or Barr bodies. The formation of sex heterochromatin takes place at an early stage of embryogenesis, after which all of the somatic daughter cells derived from the cell with inactive sex chromosome contain one X chromosome that appears euchromatic and one that is heterochromatic.

What is the composition of chromatin?

Chromatin is a working definition, and its composition can vary based on the way that it is extracted. Nevertheless, some general features persist. The relative proportion of the components of chromatin varies according to tissue, organism, and method of preparation. However, a group of proteins called histones always makes up the greatest amount of the chromosomal proteins (Table 1.1).

What are the histones?

Histones are relatively small proteins, mostly a little more than 100 amino acids in length. They are rich in the amino acids lysine and arginine. There are five classes of histones in the typical eukaryotic cell that are distinguished based on their relative content of lysine and arginine residues. The evolutionary conservation of sequence supports a structural role for histones involving comparable interactions with DNA and playing essentially the same role in all tissues and in all eukaryotic organisms (Table 1.2).

Table 1.1	Composition of Chromatin	
Component	**Percent**	
DNA	100	
Histones	114	
Non-histones	33	
RNA	7	

Table 1.2	Types of Histones			
Type	**Lys/Arg**	**Number of Ratio**	**Mass Amino Acids**	**Location (kDa)**
H1	20	215	21	linker
H2A	1.25	129	14.5	core
H2B	2.5	125	13.8	core
H3	0.72	135	15.3	core
H4	0.79	102	11.3	core

What structural role do histones play in the packing of eukaryotic DNA?

Despite the great variety and complexity of eukaryotic chromosomes, there exists a remarkable uniform structure for the structural relationship between histones and DNA. This structure is called a nucleosome.

What is the structure of the nucleosome?

Histones are complexed to the DNA in a very defined repeating pattern called the nucleosome. This basic unit of chromatin structure is roughly spherical particle of about 100 Å in diameter, which contains approximately 146 bp of coiled DNA wrapped around an intranucleosome core of two each of the four histones: H2A, H2B, H3, and H4 (Figure 1.5).

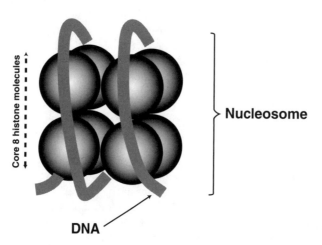

Figure 1.5   Structure of a nucleosome. The nucleosome is a nearly invariant structure in eukaryotes consisting of 146 bp of DNA, wrapped about an octamer of histone molecules.

How are nucleosomes organized?

The DNA between each nucleosome particle is called linker DNA, which represents approximately 54 base pairs. Histone 1, which is not part of the core particle, is involved in higher order chromatin structures. H1 binds the linker DNA and is believed to consolidate the nucleosome beads through head-to-tail interactions based on its asymmetric shape. This consolidation of the nucleosome beads and the coiling of the 100 Å fiber results in a higher order "solenoid" structure.

What are the higher order structures of chromatin?

The 100 Å fiber can compact further into a 200–300 Å fiber. This thick fiber is most stable and has been termed the *solenoid* structure. The 300 Å fiber is flexible and exhibits periodic discontinuities. This fiber appears to be arranged in loops that interact with a superstructure known as the nuclear matrix (or during metaphase the nuclear scaffold). These looped domains may act as structural units that also may relate to function. When chromatin is being transcribed by an RNA polymerase, the looped region is thought to uncoil into the 100 Å nucleosome units such that they would appear as "beads-on-a-string" (Figure 1.6).

Bioenergetics

What does the term *bioenergetics* mean?

Bioenergetics represents an area of thermodynamics that considers energy acquisition, exchange, and utilization in living systems.

For a given metabolic process, what determines the direction of the reaction?

Reactions may be reversible (near equilibrium) or irreversible (far from equilibrium), but the direction of the reaction to equilibrium is determined by thermodynamics and the free energy change, ΔG. The criterion for a favorable direction is that the free energy change is negative.

Why are free energy changes important to the consideration of metabolism?

The central role of free energy changes determines the favorable direction for a reaction. Thus, the metabolic pathway must be thermodynamically possible.

What is the biological significance of coupling reactions?

Unfavorable processes can be made thermodynamically favorable by coupling them to strongly favored reactions. Thus, coupling endergonic reactions to exergonic reactions is an important fundamental of metabolism and is used not only to drive the countless reactions of metabolism, but also to affect transport across membranes, transmit nerve impulses, contract muscles, and carry out other physical changes.

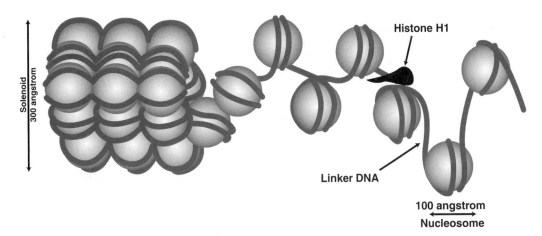

Figure 1.6 Chromatin packaging. The internucleosomal DNA, or linker DNA, is occupied by histone H1, which probably plays some role along with other proteins to condense and coil nucleosome units into a thicker fiber of approximately 300 Å (solenoid).

In coupling reactions, how does the cell use energy currency to drive the unfavorable coupled reaction?

The cell contains a variety of compounds that undergo reactions with large negative free energy changes. They include phosphate anhydrides (i.e., ATP, or Adenosine triphosphate), enol-phosphates (i.e., phosphoenol-pyruvate), some thioesters (i.e., acetyl-CoA), and compounds containing N-P bonds (i.e., creatine phosphate). Such compounds are thought of as energy currency in cells. By coupling these two reactions, the phosphorylation of glucose is now a favorable process.

$$ATP \rightleftharpoons ADP + Pi \qquad \Delta G^{0'} = -31 \text{ kJ/mol}$$
$$Glucose + Pi \rightleftharpoons Glucose\text{-}6\text{-}phosphate \qquad \Delta G^{0'} = +14 \text{ kJ/mol}$$
$$ATP + glucose \rightleftharpoons ADP + Glucose\text{-}6\text{-}phosphate \qquad \Delta G^{0'} = -17 \text{ kJ/mol}$$

Can a change in the concentration of reactants affect the direction of a reaction?

Free energy changes for a reaction are affected by the concentration of components. The equation $\Delta G = \Delta G^{0'} + Rt\ln\{[products]/[reactants]\}$ relates the free energy change to the equilibrium constant. Therefore, the relative ratio of reaction components can determine the direction of the reaction. For example,

$$Glucose\text{-}6\text{-}phosphate \rightleftharpoons Fructose\text{-}6\text{-}phosphate \qquad \Delta G^{0'} = +1.7 \text{ kJ/mol}$$

The reaction is favored in the direction of glucose-6-phosphate formation. However, in a cell, the concentration of fructose-6-phosphate can be very low because it can enter the glycolytic pathway through an irreversible step catalyzed by phosphofructokinase. As a result, the free energy change for this reaction $\Delta G = \Delta G^{0'} + Rt\ln\{[F6P]/[G6P]\}$ depends not only on the standard free energy change but also on the relative ratio of reaction components. When F6P is very low relative to G6P, the free energy change for the reaction will favor formation of F6P despite the positive value for the standard free energy change. In this way, the direction of some metabolic reactions can be influenced by the availability of either substrate or products, and a reaction appears to be unfavorable as indicated by the standard free energy change can be driven by changing the ratio of product/reactant.

How is the energy content in food measured?

The energy content in food is generally described in terms of calories. Technically, however, the dietary term *calories* actually refers to kilocalories of heat energy released by combustion of that food in the body.

What is the caloric value of food?

The caloric value of protein is 4 calories per gram. Fat is 9 calories per gram. A carbohydrate is 4 calories per gram, and for comparison, alcohol is roughly 7 calories per gram. Based on these values and the amount and composition of the food, one can roughly calculate the caloric content of food.

What factors influence energy balance?

These are dietary constituents, frequency of eating, and amounts per feeding versus one's basal metabolism, resting energy as well as nonresting energy expenditure.

What are the metabolic consequences of chronic energy imbalance where energy uptake exceeds energy expenditure?

Obesity is a metabolic disorder that results from the chronic imbalance between energy uptake and energy expenditure. Excessive adiposity is associated with dyslipidemia, elevated serum free fatty acids, cholesterol, and triacylglycerols. In addition, obese individuals most often present with hyperglycemia and insulin resistance. As a consequence, there is a direct correlation with the increased risk of co-morbidities such as cardiovascular disease, type II diabetes, and hypertension, as well as cancer.

Nucleic Acid Chemistry

Nucleotide Chemistry

What three common chemical components make up the structure of a nucleotide?

Nucleotides consist of three components: (1) a pyrimidine or purine base linked to a (2) sugar, either ribose or deoxyribose, and (3) phosphate esterified to a sugar (Figure 2.1).

How does the structure of the base moiety of nucleotides differ?

There are two types of purines, adenine (A) or guanine (G), and three types of pyrimidines, uracil (U) or cytosine (C) or thymine (T).

How are the atoms of the bases distinguished?

For purposes of nomenclature, the atoms of the rings in purine and pyrimidine bases are numbered (Figure 2.2).

Are there different types of sugars that can make up the sugar component of nucleotides?

There are two types: D-ribose or D-deoxyribose (Figure 2.3).

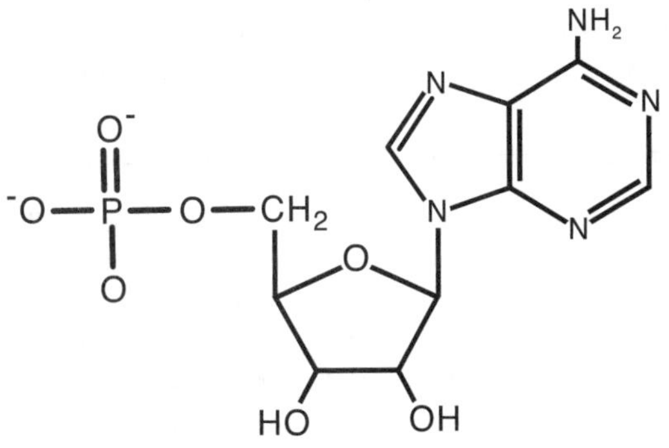

Figure 2.1 Structure of a nucleotide. The schematic illustrates the three components that make up a nucleotide; the base, which is either purine or pyrimidine, the sugar component which is either ribose or deoxyribose, and the phosphate group which is attached through an ester linkage to the sugar.

Purines   Pyrimidines

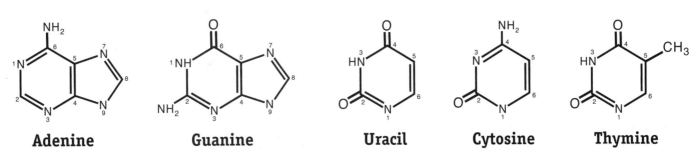

Adenine   Guanine   Uracil   Cytosine   Thymine

Figure 2.2 Numbering system for the atoms in purine and pyrimidine ring structures.

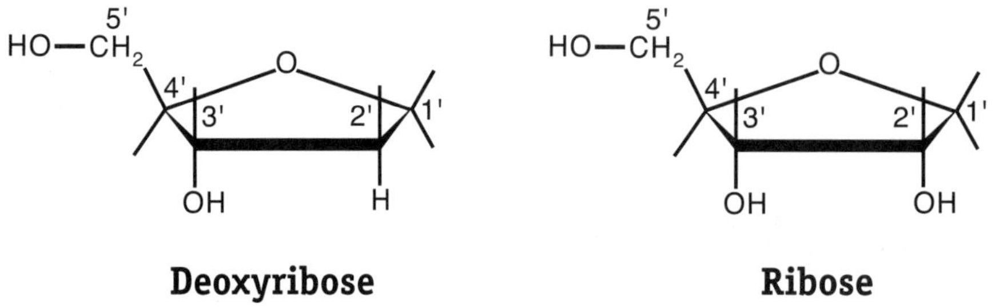

Deoxyribose   **Ribose**

Figure 2.3 Sugar components of nucleotides. Nucleic acids have two kinds of ribose sugar groups. The deoxynucleotide units of DNA contain 2-deoxy-D-ribose, and the ribonucleotide units of RNA contain D-ribose. Both types of sugars are shown in their β-furanose form.

How are the atoms of the sugar ring distinguished from those of the purine and pyrimidine rings?

The atoms in the ribose or deoxyribose units are designated by primes. This ring numbering system is important to know because the polymers of nucleotides, nucleic acids, are assigned direction (polarity) based on the ring number system of the sugar moiety of the nucleotide components.

What structural component of a nucleotide distinguishes it from a nucleoside?

Nucleotides are the phosphate esters of nucleosides. Derivatives of the nucleotides in which additional phosphates are linked as anhydrides with the phosphate groups are called nucleoside diphosphates and nucleoside triphosphates.

What component of nucleotides contributes to their ultraviolet absorption properties?

Because of the large number of conjugated double bonds in their nitrogen bases, all nucleotides absorb light strongly in the ultraviolet at approximately 260 nm. The spectrum is pH dependent because protonation or deprotonation changes the electronic distribution in the base rings. The extinction coefficient (or molar absorption properties) of the purines and pyrimidines bases is greater than that of tryptophan, the amino acid contributing most significantly to the molar extinction coefficient of proteins. When one considers that the entire polymeric structure of nucleic acids is composed of strongly absorbing monomeric units, it is certainly easy to understand why the molar extinction coefficient of nucleic acid is so much greater than that of protein.

What is the biological significance of the tautomeric forms of nucleotides?

All of the commonly occurring nucleosides and nucleotides are capable of existing in two tautomeric forms. For example, guanosine can undergo the keto–enol shift. The keto form is, however, strongly favored so that it is difficult to detect even trace amounts of the enol form. Similarly, the keto forms of thymidine or uridine are strongly preferred. Adenosine and cytidine can isomerize to imino forms, but once again, the amino forms are strongly favored. Even though the unusual tautomers are present in very small amounts, it is suggested that they contribute to the mutation process (Figure 2.4).

Are there modified forms of nucleotides?

Like amino acid components of proteins, there are also modified nucleotides in nucleic acids. This feature is particularly emphasized in the primary sequence of tRNA where N-6-methyladenine or N-7-methylguanine are examples

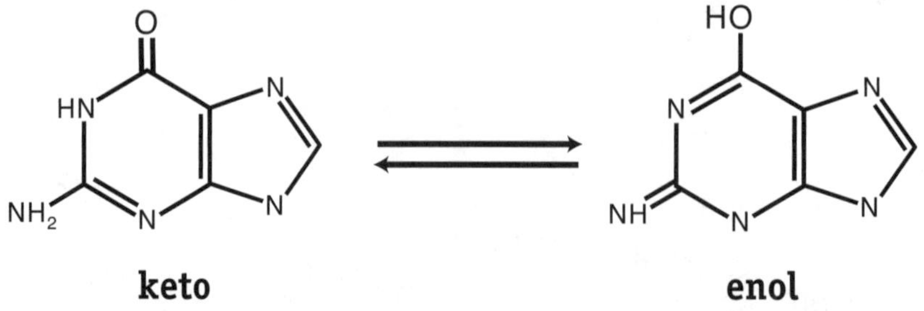

keto   **enol**

Figure 2.4 Tautomerization of the guanine ring. By shifting protons, the bases can tautomerize between amino and imino forms of A and C and keto and enol forms of G and T. The equilibria of these tautomerization reactions lie far in the direction of the amino and keto forms.

showing a modified base moiety. In contrast, 2'-O-methylcytidine is an example where the ribose moiety is modified. Modifications may also include such radical rearrangements as observed in pseudouridine. It is important to realize that these sorts of modifications, like their counterparts for modified amino acids in proteins, occur only after assembly in the macromolecule (Figure 2.5).

Biological Role of Nucleotides

What form of nucleotide represents the major energy currency of a cell?

Throughout the cell, reactions that would otherwise be unfavorable, including all biosynthetic reactions, are driven directly or indirectly by the hydrolysis of nucleoside triphosphates, particularly ATP. ATP is generated by oxidative phosphorylation and substrate level phosphorylation (Figure 2.6).

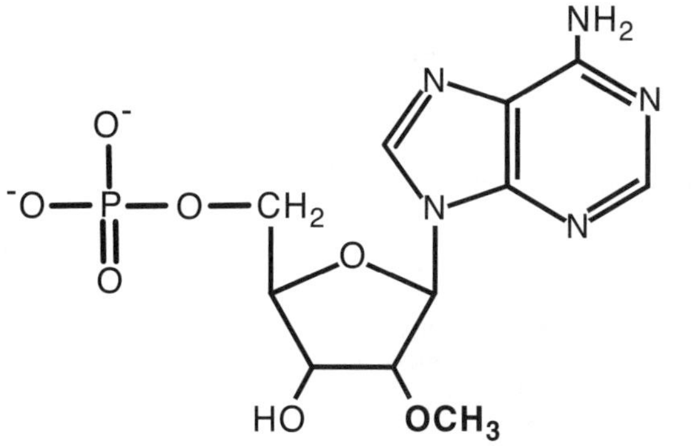

2'-O-Methyladenosine monophosphate

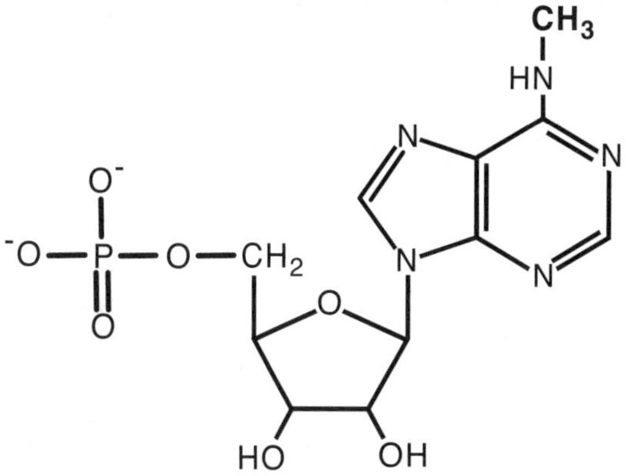

N^6-Methyladenosine monophosphate

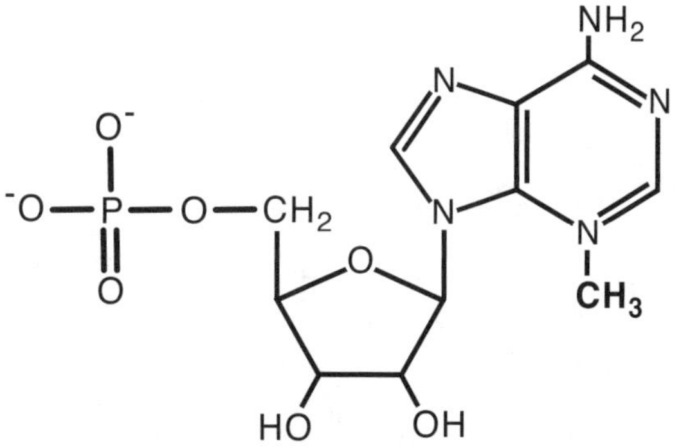

N^3-Methyladenosine monophosphate

Figure 2.5 Examples of modified nucleotides. Modifications of nucleotides most often involve the base component, but in RNA, the 2'-OH position may also be subject to modification. The modified bases are often referred to as minor bases.

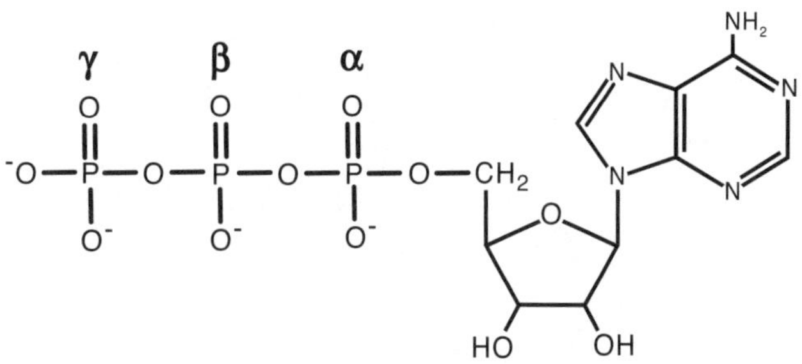

Figure 2.6 Structure of adenosine-5'-triphosphate (ATP). ATP is the major energy currency of cells. The phosphoanhydride linkages between the α and β phosphate groups and β and γ phosphate groups of ATP represent "high energy" bonds. Energy from the hydrolysis of these bonds can be used to drive biosynthetic reactions.

In what other way do nucleotides act to support synthetic reactions in the cell?

Nucleotides are involved in the activation of intermediates of metabolism such as glucose in the form of uridine diphosphate (UDP)-glucose, choline in the form of CDP-choline, methyl transfer groups in the form of S-adenosylmethionine, and sulfate groups in the form of 3'-phosphoadenosine 5'-phosphosulfate as some examples (Figure 2.7).

How may nucleotide structures be involved in catalysis?

Nucleotides are components of coenzymes such as NAD, FAD, and coenzyme A. NAD and FAD are coenzymes involved in electron transfer reactions. Coenzyme A functions as an acyl group transfer agent (Figure 2.8).

Are nucleotides involved in cell signaling responses?

Nucleotides may act as physiologic mediators such as cAMP and cGMP that are second messengers mediating responses to intracellular events (Figure 2.9).

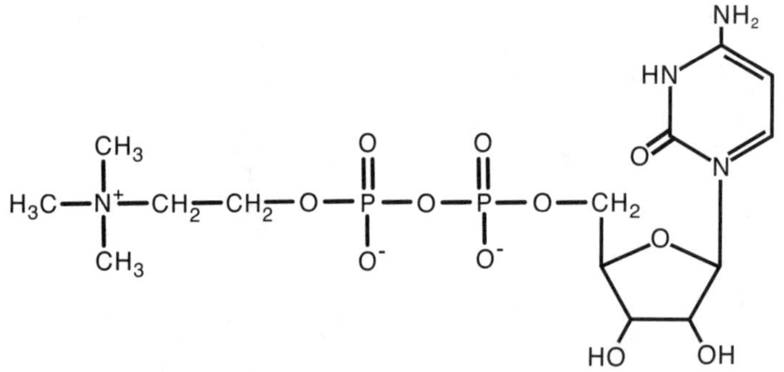

Figure 2.7   Structure of cytidine-5'-diphosphate choline (CDP-choline). Activated nucleotides intermediates are required for a variety of metabolic reactions. For example, CDP choline is utilized as an activated form of choline in phospholipid biosynthesis.

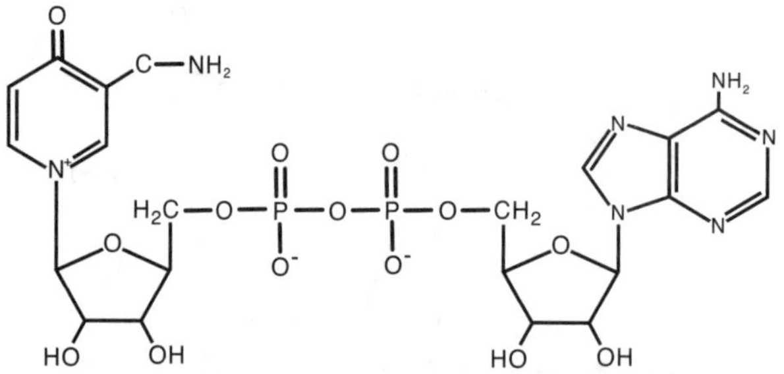

Figure 2.8   Structure of nicotinamide adenine dinucleotide (NAD⁺). Nucleotides are also found as components of coenzymes like NAD⁺ which is used redox catalyzed reactions.

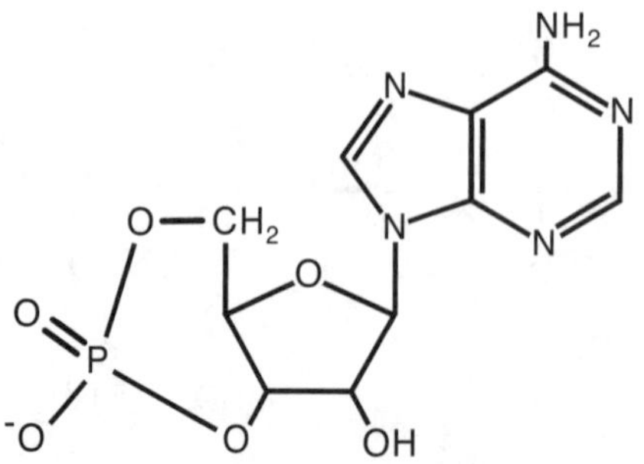

Figure 2.9   Structure of 3'-5'cyclic adenosine monophosphate (cyclic AMP). Cyclic nucleotides such as cyclic AMP serve a key role in cell signaling pathways.

In what other cellular components are nucleotides involved?

Nucleotides serve as the monomeric units of nucleic acids and are therefore used in the assembly of DNA and RNA.

Polynucleotide Chemistry

What defines the primary structure of polynucleotides?

Again, like its counterpart the protein molecule, nucleic acids are polymers. The polymeric structure of nucleic acids has a defined primary structure relating to the sequence of purine and pyrimidine nucleotides linked together through the 5'-phosphate of one nucleotide to the 3'-hydroxyl group of the sugar moiety of the adjacent nucleotide.

What is the nature of the linkage that forms the polymeric structure of nucleic acids?

Phosphodiester linkage refers to the bridge between each successive nucleotide of a polynucleotide and is to nucleic acid chemistry as the peptide linkage was to protein chemistry. A polynucleotide is then a long, unbranched polymer formed by bridges between the 5'-phosphate and the 3'-hydroxyl groups (Figure 2.10).

How does the term *polarity* relate to the structure of a polynucleotide chain?

One important feature of the polynucleotide chain is that it has direction, or what nucleic acid chemists call polarity. This relates to the sugar moiety of the nucleotide. This means that their ends (or termini) are not the same. The 5'-end refers to the 5'-carbon of the sugar moiety, and the 3'-end similarly refers to the 3'-carbon of the sugar moiety not linked in a phosphodiaster band.

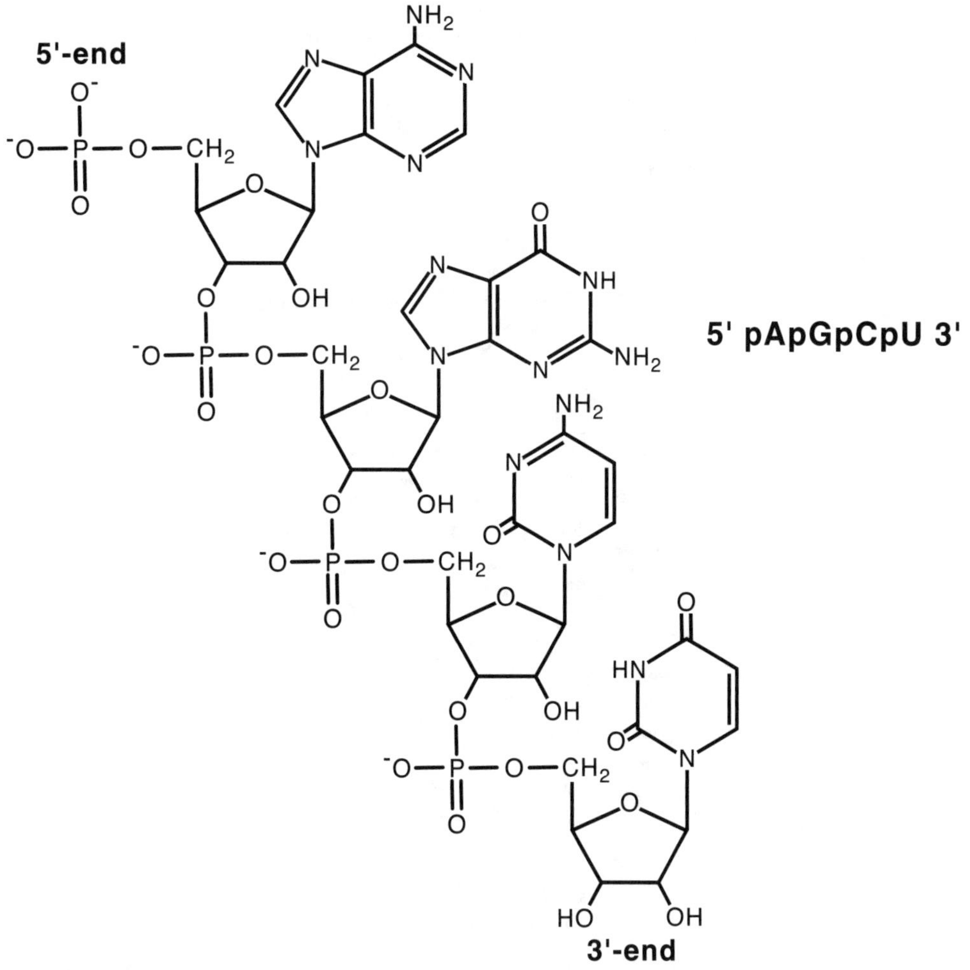

Figure 2.10 Structure of a polyribonucleotide chain. The phosphodiester bond links successive nucleotide units in a polyribonucleotide chain. The backbone of the polynucleotide chain has direction which is called polarity. Polarity relates to the position of the ribose moiety which is not linked in a phosphodiester bond and creates the end of the polynucleotide chain.

How is a sequence of nucleotides making up a polynucleotide chain written?

By convention, the sequence of all polynucleotide chains are written in the 5'→3' direction. That is, with the 5'-end to your left and the 3'-end to the right. As a corollary, it should be noted that AGCGUG and GUGCGA are different polynucleotides.

Secondary Structure of Polynucleotides

What are the salient features of the Watson and Crick model for structure of DNA?

The DNA molecule is double stranded and in the form of a right-handed helix with the two polynucleotide chains wound round the same axis and held together by hydrophobic interactions and hydrogen bonds between the bases (Figure 2.11).

How did this model accommodate Chargaff's finding of the equivalence of adenine and thymine and guanine and cytosine in most naturally occurring DNA molecules?

By making scale models, Watson and Crick were able to show that the bases could fit in the interior if they were arranged in pairs of one pyrimidine and one purine, where the adenine base on one strand would pair through hydrogen bonding with thymine base on the opposite strand, and guanine would pair through hydrogen bonding with cytosine on the opposite strand.

What were the most important consequences of the base-pairing configuration found in DNA?

The most important consequence was that the order in which the bases occur in one chain automatically determines the order in which they occur in the other complementary chain of the double helix. Apart from this essential condition, there are no restrictions on the sequences of the bases along the chains. Thus, the secondary structure in nucleic acids (also in RNA) involves hydrogen bonding between base pairs.

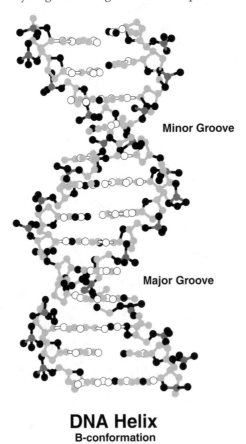

Minor Groove

Major Groove

DNA Helix
B-conformation

Figure 2.11   Right-handed, double helical structure of DNA. The original model proposed by Watson and Crick had ~10 base pairs per helical turn, where the two chains rap around each other forming a right-handed helix. The hydrogen bonding of the bases between the two chains of the helix was strictly based on A to T and G to C base pairing, which also help stabilize the secondary structure. The two chains of the helix have opposite polarity.

How do these Watson-Crick base pairs relate to genetics?

The specificity in base pairing caused by hydrogen bonding is the basis of the genetic code, and these specific base pairs are often referred to as the Watson-Crick base pairs.

What are the most important features of DNA structure?

(1) The order that the bases occur in one chain automatically determines the order in which they occur in the other complementary chain. (2) There are no restrictions on the sequence of the bases along the chains. (3) The pairs of bases are flat and may be stacked one above the other like a pile of plates (hydrophobic interactions) so that the molecule is radially represented as a spiral staircase with the base pairs forming the steps. (3) The distance between the sugar–phosphate backbone of each base pair is essentially identical. (4) The fact that G:C base pairs involve one more hydrogen bond makes this more stable than A:T. (5) Any alteration in the specific functional groups involved in base pairing can alter the specificity of these interactions. (6) The two polynucleotide chains are of opposite polarity in the sense that the internucleotide linkage in one strand is $3' \rightarrow 5'$, whereas in the other it is $5' \rightarrow 3'$. (7) The two helices are right handed and cannot be separated without unwinding.

How does DNA convey biological information?

DNA encodes biological information in at least two different ways. First, through the linear order of nucleotides, it specifies specific sequences that dictate the boundary of transcribed regions and regulate their transcription. Second, through its shape, DNA can provide information that is used directly or indirectly by a variety of macromolecules to regulate access to DNA sequence information.

What forces determine the conformation of DNA?

The hydrophobic properties of the bases are, to a large extent, responsible for forcing polynucleotides to adopt helical conformations. The bases contain polar groups at their planar edges that may interact with H_2O. The faces, however, are unable to participate in such interactions and tend to avoid contact with H_2O. Rather, they tend to interact with one another. The stability of this arrangement is further reinforced by an interchange between electrons that circulate in the *pi* orbitals located above and below the plane of each base ring. This interchange between adjacent base-ring structures is known as stacking interactions.

What other shapes (conformations) of DNA can form?

DNA fibers can form other structures with the B structure or conformation corresponding to the original Watson-Crick model. The A and C structures are also right-handed double helices but differ in the pitch and in the number of bases per turn. In addition, the bases are not flat but are tilted in both A and C conformations. Their biological significance is not clear, but the A conformation is believed to be very close to the structure adopted by double-stranded RNA and DNA–RNA hybrids. A key finding is that these and other local variations of the double helix depend on base sequence. A protein searching for a specific target sequence in DNA may sense a conformational change through its effect on the precise shape of the double helix.

Why do DNA–RNA duplexes and RNA–RNA duplexes structures take on the A conformation?

Because of the presence of the extra 2'-hydroxyl group, RNA is unable to adopt the B conformation. Thus, when it is engaged as a template for making RNA, the DNA molecule must adopt the A conformation in that region.

Does RNA possess secondary structure?

Although RNA molecules do not possess the regular interstrand hydrogen-bonded structure of DNA, they have the capacity to form double-helical regions. These helices can be formed between two separate RNA chains, but are more frequently found between two segments of the same chain folded back on itself. The secondary structure is similar to the A form of DNA with tilted bases because the 2'-OH hinders B-conformation formation. The helical regions formed in this manner are seldom regular because the segments on the chain brought into opposition do not have entirely complementary sequences; thus, nonbonded residues loop out of the structure. In some RNA molecules, such as tRNA, as much as 70% of the bases are involved in secondary structure interactions (Figure 2.12).

Can nucleic acid molecules with secondary structure lose this structure through denaturation?

Denaturation of nucleic acids involves conditions that disrupt the two types of forces that stabilize secondary structure in nucleic acids, hydrogen bonding, and hydrophobic interactions. Conditions such as organic solvents, ionic strength, pH, and temperature (thermal denaturation) all can be used to disrupt the secondary structure of nucleic acids.

Can the loss of secondary structure of DNA be monitored?

When a solution of DNA is heated, very striking changes occur in many of its physical properties, such as viscosity, light scattering, and optical properties. This change occurs in a narrow temperature range, which depends on the pH,

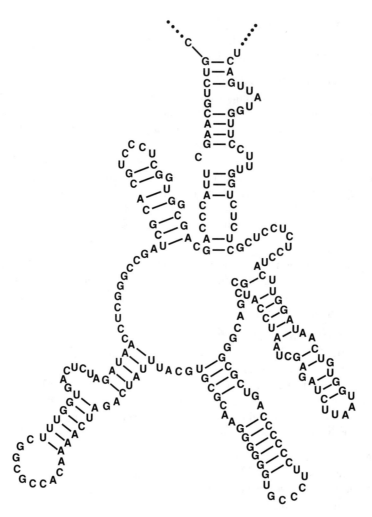

Figure 2.12 Secondary structure found in RNA. The potential for base-paired helical structures in RNA may be extensive as illustrated for a segment of RNA shown in this figure.

type of buffers, and ionic strength, and for most DNA occurs in the region of 80°C to 90°C, but may be as little as 40°C in the presence of reagents such as formamide, which assist denaturation. The denaturation or melting of DNA is usually characterized by the temperature of midpoint transition, T_m, sometimes referred to as the melting temperature. This process represents the disruption of the double helix, loss of the duplex structure.

Why does the extinction coefficient (or molar absorption properties) of DNA change on thermodenaturation?

The heterocyclic rings of the bases absorb ultraviolet light strongly, with a maximum around 260 nm, but the absorption of DNA itself is approximately 40% less than would be expected from a mixture of free nucleotides of the same composition. The loss of absorption properties as a result of secondary structure is called the "hypochromic effect." This phenomenon appears to result from mutual interactions of electron systems of the bases. The degree of hypochromicity is a sensitive measure of the physical state of DNA because any departure from the ordered configuration of the double helix is reflected by a loss of hypochromicity.

Why is melting temperature, T_m, directly proportional to the G:C content of the DNA?

The G:C base pair contains three hydrogen bonds instead of the two contained by the A:T base pair. DNA with a higher G:C content would, therefore, be expected to have greater secondary structure stability. Base stacking is also believed to contribute to the stability of the duplex structure. In general, the interaction energy gained by stacking between adjacent G:C base pairs is greater than that gained by interaction between A:T base pairs.

What are the most general features that distinguish DNA from RNA?

Cells contain two types of nucleic acids that can be distinguished by the sugar moiety, ribose (RNA) versus deoxyribose (DNA), by a pyrimidine base where DNA contains the base thymine, whereas RNA contains uracil, and by structure where DNA exists primarily as a double-stranded structure, whereas RNA is considered, for the most part, as single-stranded.

How does the variation in sugar moieties between the two types of nucleic acids affect the chemistry of the two nucleic acid molecules?

DNA is stable in alkali, whereas RNA is hydrolyzed. This is due to the activation of the 2'-OH group as a nucleophile that attacks the adjacent phosphodiester bond. The presence or absence of the 2'-OH group ultimately affects the possible secondary structure of the nucleic acid. The difference in sugar moiety becomes a specific recognition handle for enzymatic sensitivity and therefore selectivity. Differences in sugar groups permit the differential quantitation of the two types of nucleic acids in tissue extracts.

What chemical conditions lead to the hydrolysis of nucleic acids?

Dilute acids degrade DNA more rapidly than RNA, but in both cases, the degradation occurs by depurination (loss of adenine and guanine) producing apurinic sites (breakage of the N-glycosidic bond between the sugar and base moiety of a nucleotide) and subsequent hydrolysis of the phosphodiester bond. Hydroxide ions hydrolyze RNA to 2'-and 3'-mononucleotides. In contrast, the phosphodiester bond in DNA is resistant to alkaline hydrolysis, and under alkaline conditions, DNA only loses its secondary structure.

Are there enzymes that degrade nucleic acids?

Nucleases are a general class of enzymes that degrade nucleic acid. Nucleases may be characterized, like other depolymerizing enzymes, by their substrate specificity. For example, an endonuclease does not require recognition of the ends of the nucleic acid molecule, hydrolyzing internal bonds, whereas an exonuclease requires the recognition of a terminus before initiating attack. Nucleases may show specificity for either a double-stranded or a single-stranded structure, base or sequence, DNA or RNA, or be nonspecific.

What are restriction endonucleases?

Restriction endonucleases are strain-specific enzymes that enable bacteria to recognize and rapidly destroy foreign DNA by introducing double-stranded scission at a limited number of sites, rendering the DNA susceptible to total hydrolysis. The restriction endonucleases are named accordingly. For example, consider restriction endonuclease EcoRI: the first three letters give the name of the bacteria, such as *Eco* = *E. coli*; and R refers to the resistance factor plasmid that carries the gene that encodes this specific restriction endonuclease, and I refers to the existence of more than one restriction enzyme coded by this plasmid.

What is a common feature of the specific sequence recognized by restriction endonucleases?

An important characteristic of the restriction endonuclease is that they cleave DNA only within, or in some instances near, specific nucleotide sequences known as palindromes. A particular restriction nuclease will cut the long length of a DNA double helix into a series of fragments known as restriction fragments.

Regarding restriction endonuclease, what are isoschizomers?

Isoschizomers are restriction endonucleases that recognize identical nucleotide sequences.

$$\textit{Hae}\text{II} \quad \text{GG}\downarrow\text{CC} \qquad\qquad \textit{Hpa}\text{II} \quad \text{C}\downarrow\text{CGG}$$
$$\text{CC}\uparrow\text{GG} \qquad\qquad\qquad \text{GGC}\uparrow\text{C}$$

However, as shown by the arrows (cleavage site), they do not necessarily cut DNA the same way.

3 DNA Metabolism

Replication

What types of enzymes synthesize DNA?

Two types of enzymes synthesize DNA. The difference between these two groups of enzymes is based on the nature of the template copied. DNA-dependent DNA polymerases, most often referred to as DNA polymerases, copy a DNA template to synthesize a complementary DNA strand and, therefore, are essential for DNA replication. On the other hand, RNA-dependent DNA polymerases, which are referred to as reverse transcriptases, copy an RNA template to synthesize a complementary DNA strand.

How does the catalytic efficiency of a DNA polymerase relate to synthesis of DNA?

Catalytic efficiency or the processivity of the DNA polymerase relates to the capacity of the DNA polymerase to remain associated with the template strand and to move without delay to the newly generated primer end. Because dissociation DNA polymerase and its reassociation with the primer terminus can consume a minute of time compared with the millisecond required to add a nucleotide, catalytic efficiency or processivity has a major effect on the overall rate of replication. For example, in *E. coli*, DNA polymerase I, with a processivity of 10 to 20 nucleotides, is better suited for filling in short gaps in DNA repair, whereas DNA polymerase III holoenzyme, responsible for most replication and equipped to clamp itself to the template, has virtually unlimited processivity. Thus, replication of a nucleotide chain that takes the DNA polymerase III holoenzyme only 10 seconds to replicate would take DNA polymerase I several hours.

How does the property of fidelity relate to the enzymatic synthesis of DNA?

Fidelity of a DNA polymerase describes the accuracy in copying a nucleotide sequence of the template strand.

How is chromosome replication fidelity maintained?

Chromosome replication fidelity is generally considered to be determined by a combination of base selection and error selection activities of DNA polymerases along with postreplication mismatch repair.

What error selection activities do DNA polymerases possess?

Initially, a deoxyribonucleotide is added to the 3'-hydroxyl end of the primer DNA chain to pair with the base next in line on the template chain. The error rate may be in the range of 1 in 10,000 events (the error rate is approximately 10^{-4} per base replicated). If the DNA polymerase possesses or has an associated proofreading activity, a $3' \rightarrow 5'$ exonuclease, then this activity coordinately removes a newly added mismatched nucleotide, decreasing the error rate to approximately 1 in 1,000,000 events (an error rate of approximately 10^{-6}).

Why are there different DNA polymerases in cells?

Inherent properties of the DNA polymerase and accessory proteins contribute to their specialization and capacity to participate in replication of DNA associated with S phase, DNA replication associated with DNA repair, and DNA replication associated with DNA recombination.

How many DNA polymerases are typically found in eukaryotic cells?

In humans, eight (and most recently possibly nine) DNA polymerases have been identified. All DNA polymerases studied thus far from any biological source synthesize DNA by the Watson-Crick base pairing rule, incorporating A, G, C, and T opposite the templates T, C, G, and A, respectively. However, their fidelity may vary considerably.

What is the role of each of the known DNA polymerases?

Nuclear DNA replication involves DNA polymerases α, δ, and ϵ. Mitochondrial DNA replication requires DNA polymerase γ. Excision repair synthesis appears to involve DNA polymerase β. The DNA-induced mutagenesis pathway for translesion DNA synthesis is believed to involve DNA polymerase ζ. DNA lesion bypass polymerase capable of both error-free and error-prone translesion synthesis is believed to involve DNA polymerase η. Repair of interstrand crosslinks is believed to involve DNA polymerase θ. DNA polymerase ι appears to function in the DNA synthesis that results in somatic hypermutation during immunoglobulin development. These role assignments are not meant to be exclusive, and some assignments are tentative because of redundancy in the system.

How does the model of the replication fork account for the apparent same direction of synthesis of the two parental template strands that run in opposite direction?

Because any DNA polymerase can synthesize DNA only in the 5'→3' direction, the model for the replication allows for continuous synthesis on the template strand with 3'→5' polarity (leading strand). However, for the template strand with 5'→3' polarity (lagging strand), synthesis is discontinuous, producing short, newly synthesized DNA strands ("Okazaki fragments"), which are subsequently processed and joined together. This model allows DNA synthesis on both template strands at the replication fork to appear to progress in the same direction relative to fork movement (Figure 3.1).

Which DNA polymerase in eukaryotes is involved in discontinuous strand ("Okazaki fragment" synthesis) on the lagging strand?

DNA polymerase α (together with primase) is involved in "Okazaki fragment" synthesis. After removal of the RNA primer, the extension of these fragments and formation of a continuous newly synthesized daughter strand may involve either DNA polymerase δ and/or ϵ.

Which of the DNA polymerases in eukaryotes is involved in leading strand synthesis?

DNA polymerase δ is required for leading strand synthesis. Proliferating cell nuclear antigen has been shown to be a cofactor of DNA polymerase δ that acts as a "clamp" to increase greatly the processivity of DNA polymerase δ.

What is primase, and why is it required for DNA replication?

Because a DNA polymerase can only extend a pre-existing chain, a new chain is initiated by a short RNA transcript, known as a primer. Primase is responsible for synthesizing the short RNA primer.

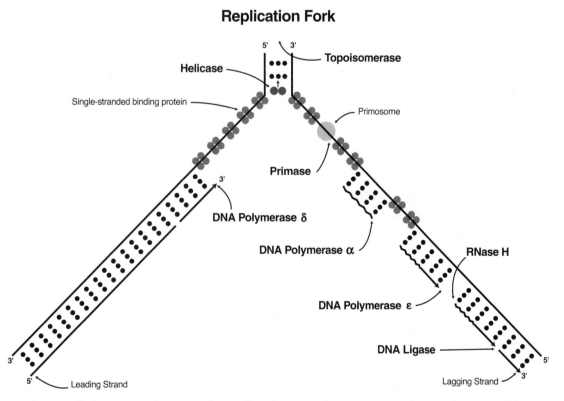

Replication Fork

Figure 3.1 Replication fork. DNA is always synthesized in the 5'→3' direction. Thus the template strand is read in the 3'→5' direction. To accommodate the direction of DNA synthesis, newly synthesized DNA is copied from the leading strand in a continuous fashion; whereas, newly synthesized DNA is copied discontinuously on the lagging strand in short pieces, in a direction opposite to the direction of replication fork movement.

Is the progression of the replication fork controlled?

Progression of the replication fork is essentially controlled by concurrent replication of both strands, rather than the jerky sequence of synthesis of one strand and then the other. Control seems to be exerted by having the priming of nascent fragments (Okazaki fragments) of the lagging strand integrated with continuous synthesis of the leading strand.

How are parental template strands of DNA made available for copying by the DNA polymerases?

To facilitate parental strand separation, cells contain a class of specific proteins called helicases. Helicases use energy derived from the breakdown of ATP to move progressively along and to disrupt the parental helix.

Why are single-stranded binding proteins critical to the progression of the replication fork?

A newly created single-stranded region does not remain free after helicase moves along the double helix. To prevent reassociation, the single strands of parental DNA are quickly covered by specific single-stranded DNA-binding proteins (designated SSB). Unlike helicases, SSBs do not require ATP and have no known enzymatic function.

What function does RNase H-type activity serve in DNA replication?

RNase H (H standing for hybrid, the RNA–DNA hybrid) activity removes the RNA primer from newly synthesized fragments (Okazaki fragments), creating a gap. This gap between adjacent Okazaki fragments is filled in by a DNA polymerase, and the fragments are joined by DNA ligase.

How is the continuous strand of newly synthesized DNA formed from the discontinuously synthesized fragments during lagging strand synthesis?

DNA ligase joins the 3'-OH group to the 3'-phosphate of the adjacent fragment based paired to the template-lagging strand. DNA ligase requires energy for ligation by coupling ligation with hydrolysis of either ATP in most cells, including eukaryotes, or nicotinamide adenine dinucleotide (NAD) in bacteria.

Because DNA is a helix with the parental strands wound around each other, the separation of the parental strands by DNA helicases produces a problem of the overwinding of DNA in front of the replication fork. How is this problem overcome to allow progression of the replication fork?

As replication of the two daughter strands proceeds along the helix, the overwinding (positive superhelical turns) is overcome by enzymes called topoisomerases. These enzymes produce a variety of topologic changes in DNA to remove the superhelicity generated by the separation of the parental template strands and the progression of the fork.

How are the ends of the linear DNA molecules of chromosomes maintained?

A eukaryotic chromosome contains a linear DNA molecule and hence a question of how termination occurs. Left without any special replication machinery, the DNA in eukaryotic chromosomes would become shorter after each round of replication if there were no means to compensate for the inability to complete synthesis on the lagging strand at each end of the chromosome. This problem is solved by the unique structure of telomeres, the ends of chromosomes, and an activity called telomerase that specifically adds repeat sequences that can self-anneal, allowing for the extension and filling in of the unreplicated ends.

How is DNA replication initiated?

Unlike most bacteria, eukaryotes initiate replication at multiple sites called *origins of replication* within the chromosome (Figure 3.2).

What is the origin of replication?

The term *origin of replication* is used to describe the DNA sequences at which replication initiates as well as *cis*-acting sequences that promote initiation. When the cell enters S phase, a nucleoprotein complex is formed between the initiator, a replicon-specific protein, and the replicator, a DNA sequence that usually contains repeated elements important in recognition and binding by the initiator protein. It is inferred that the initiator wraps the DNA into a three-dimensional structure that serves as a nucleus for a series of protein–DNA and protein–protein interactions that lead to the initial opening of the helix, priming and DNA synthesis by a DNA polymerase–primase complex.

Are all DNA sequences replicated at the same time?

Multiple sites of replication are initiated on eukaryotic chromosomes; each site is considered a replicating unit or "replicon." Replication of euchromatic material proceeds in the first part of S phase, whereas heterochromatic material is replicated in late S phase. The length of time in S phase appears to be dependent on the number and efficiency of replicon initiation.

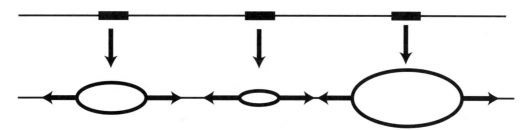

**Multiple sites of replication are initiated
on eukaryotic chromosomes.**

Figure 3.2 Multiple sites of replication on a eukaryotic chromosome. In order to replicate the amount of DNA in eukaryotes during S phase of the cell cycle, initiation of DNA synthesis occurs at multiple sites on each chromosome.

How is DNA replication regulated?

Eukaryotic DNA replication is tightly controlled by a range of complex mechanisms. DNA synthesis normally initiates at many specific origins of replication on each chromosome, probably in response to activation by cyclin-dependent kinases. Different origins may initiate at different times, and replication forks are spatially organized into clusters within the nucleus. Replication produces exactly one copy and normally avoids reinitiation within a single cell cycle. Progress of DNA replication is also monitored so that progression to subsequent cell-cycle phases can be delayed pending completion of S phase.

DNA Repair

How significant is the ability to repair DNA to human cells?

Usually cells eliminate DNA damage by molecular DNA repair mechanisms. Because the properties of cells and the organisms constructed from them ultimately depend on the DNA sequences of their genes, irreversible alterations in a few DNA base pairs can cause substantial changes in the corresponding organism. The inability to repair DNA damage appropriately ultimately causes cancer in humans.

How do changes in DNA sequences effect the information in DNA?

Because of the central role of DNA, damage to DNA has more severe implications for the functional integrity of the cell than damage to most other cellular components. This damage may lead to effects such as inhibition of DNA replication or transcription and misinformation at the level of proteins derived from affected genes.

What is a mutation?

A mutation is a stable change in the DNA sequence of a gene. The changes in DNA sequence referred to as mutations may be hidden or visible (phenotypically expressed).

What types of damage lead to changes in DNA sequence?

Single-base alterations can be caused by deamination or the alkylation of bases. Two-base alterations can be caused by ultraviolet light that leads to the dimerization of adjacent pyrimidine residues on one strand of the DNA helix. Chain breaks can be caused by some types of alkylations, but are particularly observed with x-rays and radiation that produce free radicals. The free radicals react with the base structure of nucleotides in DNA to produce base peroxides. These are chemically unstable, which leads to base loss and strand breakage. Some alkylating agents that are bifunctional, such as the nitrogen mustards or mitomycin C, may even produce cross-links between DNA strands of the helix.

Are there different pathways of DNA repair?

Human cells are endowed with many different pathways for the repair of DNA, and corrective processes are also an integral part of DNA replication and transcription.

What general pathways exist for the repair of single-strand damage in DNA?

Two general pathways, previously referred to as "light repair" and "dark repair," have been extensively studied for the repair of single-strand damage in DNA. Although extensively studied in bacteria, the "light repair" pathway, also known as "photoreactivation pathway," does not appear to be applicable to humans. "Dark repair" refers to excision repair pathways.

What advantage does the photoreactivation pathway serve in bacteria?

Enzymatic photoreactivation takes advantage of a specific enzyme called photolyase. Photolyase repairs ultraviolet light–induced damage in DNA directly by monomerizing pyrimidine dimers in a visible light-dependent reaction. This is an efficient and simple repair pathway, relying on only one enzyme, photolyase, and it does not require the presence of a complementary undamaged DNA strand. This repair system acts specifically on pyrimidine dimers, the principle lesions caused by ultraviolet light.

$$\text{DNA} \xrightarrow[\text{(260 nm)}]{\text{Ultraviolet light}} \underset{\text{(damaged DNA)}}{\text{DNA-dimers}} \xrightarrow[\text{(310–440 nm)}]{\substack{\textit{Photolyase+}\\ \text{Visible light}}} \text{Repaired DNA}$$

What pathways carry out excision repair (dark repair)?

Excision repair comprises a number of pathways that repair an almost infinite variety of DNA lesions by sequential steps involving the degradation of damaged segments, their resynthesis using undamaged complementary strand as template, and the ligation of the new patches of DNA to the old undamaged segments. These pathways may be divided into two groups according to whether repair is initiated by incision of the damaged DNA strand (nucleotide excision repair) or by the removal of a damaged base (base excision repair).

How does nucleotide excision repair remove damaged DNA?

Nucleotide excision repair involves an enzyme system that hydrolyzes two phosphodiester bonds, one on either side of the lesion, to generate an oligonucleotide carrying the damage. The excised oligonucleotide is released from the duplex, and the resulting gap is then filled in and ligated to complete the repair reaction.

What is the specificity of the nuclease activity associated with nucleotide excision repair?

In general, the incision pattern is rather precise, and as a consequence, depending on whether the lesion is a monoadduct or a diadduct, the damage is removed in 12 to 13 nucleotide oligomers in prokaryotes and in 27 to 29 nucleotide oligomers in eukaryotes. The nuclease activity, which is unique to DNA, has been named excinuclease (excision nuclease) to differentiate it clearly from endonucleases and exonucleases that perform other functions in the cell. The excinuclease is an operational definition of a protein complex that binds DNA and uses the energy of ATP to deform DNA and to excise the damage by dual incisions.

What types of damage does the excinuclease repair?

In humans and in *E. coli,* the excinuclease is the sole enzyme system for removing bulky DNA adducts. These include the carinogenic cyclobutane pyrimidine dimers induced by ultraviolet light, the benzo[α]pyrene–guanine adducts caused by smoking, the thymine–psoralen adducts and the guanine–cisplatin adducts caused by chemotherapeutic drugs. The enzyme also repairs many other lesions that do not distort the helix, including O^6-methylguanine and other methylated bases.

Can the excinuclease repair mismatched nucleotides?

The excinuclease system even "excises" mismatched nucleotides from DNA. However, in contrast to the mismatch repair system that has a built-in mechanism enabling it to differentiate the correct strand from the wrong strand, the excinuclease excises the mismatched base from either strand and may actually cause mutation fixation rather than mutation avoidance.

How does base excision repair differ from nucleotide excision repair?

The base excision repair pathways are initiated by a family of enzymes called DNA glycosylases (or DNA N-glycosidases or DNA glycohydrolases). These enzymes remove damaged bases by cleaving the bond between the base and deoxyribose without breaking the DNA strand. Subsequently, the base-free site is recognized by an AP (apurinic/apyrimidinic) endonuclease that cleaves the DNA strand 5' to the damage. This allows for the degradation of the damaged segment and resynthesis to take place. Base excision repair has a limited substrate range because the DNA glycosylases that initiate the repair process are in intimate contact with the lesion during catalysis.

Does a base excision repair pathway remove uracil residues that may be misincorporated during the replication of DNA?

A member of the family of enzymes of the base excision repair pathways called uracil N-glycosylase specifically recognizes uracil residues in DNA and catalyzes the removal of uracil by cleaving the bond between uracil and deoxyribose to produce a base-free site.

Do some base excision glycosylases show a greater range of specificity?

In human cells, for example, the 3-methyl adenine-DNA glycosylase catalyzes the excision of a broad variety of modified bases, including N-methylpurines generated by alkylating agents, deamination products such as hypoxanthine, and 1, N^6-ethenoadenine, an adduct generated by chloroacetaldehyde or products of lipid peroxidation.

What repair mechanism is responsible for the correction of errors generated during DNA replication?

The postreplication mismatch repair system is responsible for the correction of errors generated during replication that have escaped DNA polymerase proofreading.

How does the mismatch repair system know whether to repair the parental or newly replicated DNA strand?

Based on a comparison to the prokaryotic example, *E. coli*, mismatch recognition steps include (1) mismatch recognition by proteins similar to MutS and MutL, directed by methylation patterns of the template strand that allow newly synthesized strand discrimination; (2) incision by a MutH endonuclease counterpart in the unmodified (newly synthesized strand) at sites opposite the modified (methylated sequence); (3) degradation from a nick toward the mismatched site by exonuclease; and (4) gap filling by DNA polymerase followed by DNA ligation. In eukaryotes, it may be the patterns of 5-methylcytosine in CpG sequences that designate the parental strand.

What rare autosomal recessive disease, clinically characterized by hypersensitivity to sunlight, contributed to the understanding of the mechanism of nucleotide excision repair?

Xeroderma pigmentosum (XP) is a rare autosomal recessive disease clinically characterized by hypersensitivity to sunlight, abnormal pigmentation, and predisposition to skin cancers, especially on sun-exposed areas, caused by genetic defects in nucleotide excision repair.

What aspect of nucleotide excision repair is compromised in XP patient's cells?

Cell fusion studies have revealed the presence of a number of genes encoding the XPA, XPB, XPC, XPD, XPF, and XPG proteins that are involved in NER. Each encodes a protein and participates in the assembly of the excinuclease complex, along with other proteins, that recognizes and excises damaged sequence in DNA.

How does Cockayne syndrome differ from XP, as both involve genetic lesions in nucleotide excision repair?

Cockayne syndrome (CS) is also known as a repair-deficient human disease characteristic of postnatal failure of growth, a limited life span, and progressive neurologic dysfunction. CS has two different genes that have been identified. These are distinct from those of XP and when affected result in tissue that is moderately sensitive to ultraviolet irradiation. A defect in either of these two genes associated with CS affects a subpathway for nucleotide excision repair involving a transcription-coupled repair process capable of removing particular lesions from transcribed strands of active genes.

DNA Recombination

What is DNA recombination?

DNA recombination is a process that results in the formation of new arrangements of genes by the movement of segments of DNA.

When does DNA recombination take place?

In higher eukaryotes generally and in mammalian cells in particular, DNA recombination occurs rarely, if at all, outside of the meiotic process. This means that in higher animals, when cells of the germ line differentiate into spermatocytes or oocytes, an activation of the recombination process occurs. In fact, a major role of meiosis is to allow recombination between homologous chromosomes. Normally, this exchange is precise. However, on rare occasions, nonreciprocal exchange can occur, resulting in deletion of information.

What is the function of meiotic recombination?

The formation of new linkage of genes leads to genetic diversity. Thus, meiotic recombination provides a genetic mixing to ensure that each offspring is more than a 50–50 mix of parental genes.

In addition to meiotic recombination, how else is genetic diversity maintained in a species?

Genetic diversity is maintained through both mutation, which alters single genes or small groups of genes in an individual, and recombination, which redistributes the contents of a genome.

Does DNA recombination ever occur outside of meiosis?

Perhaps the most noticeable and important example of recombination that occurs outside of meiosis in somatic cells is the process that occurs during the normal differentiation of plasma cells to antibody-producing cells. For example, during embryonic development, a recombination event relocates gene segments, which include information coding for the variable region of either the light or heavy chains of immunoglobulins to a position essentially adjacent to the constant coding region. As a result, new functional genes are formed in a particular population of cells.

Are there basic features of the mechanism by which DNA segments recombine?

At present, a complete metabolic interpretation of the process of DNA recombination in mammalian tissue is not understood. However, a prototype model that uses a recombination intermediate envisioned by Robin Holliday has been generally accepted.

What are the essential features of the Holliday recombination intermediate?

The Holliday model proposes that nicks in the recombining chromosomes promote a strand exchange reaction after two homologous double helices align. In each, the same strand is nicked, and the free ends created leave the complementary strands to which they had been hydrogen bonded and become associated instead with the complementary strands in the homology double helix. The result of the reciprocal strand exchange is to establish a transient physical connection between the two DNA molecules that are going to recombine.

Are there specific proteins involved in DNA recombination?

As a model, the bacterial system suggests that there is a specific group of proteins that is required for DNA recombination. There are proteins that recognize homologous sequences on paired chromosomes to facilitate strand exchange. There are also enzymes that catalyze the nicking of the DNA and the traditional replication machinery that repairs these regions after recombination.

How does transposition differ from homologous DNA recombination?

Transposition is the movement of a gene from one chromosome to another or from one site to a different site on the same chromosome. In contrast to homologous DNA recombination, transposition does not require extensive homology of nucleotide sequence. The human genome contains a variety of mobile sequences. Such transposable elements range in length from a few hundred to tens of thousands of base pairs, and they are often present in multiple copies per cell.

Do these transposable elements move frequently about the chromosome or chromosomes?

Movement of transposable elements in the human genome is a very rare event. For this reason, it is often difficult to discriminate transposable elements from nonmobile parts of chromosomes.

Are there ways that DNA can recombine other than by general recombination or by transposition?

Another mechanism by which segments of DNA recombine is called "site-specific" recombination. This involves the exchange of two, but not necessarily homologous, DNA sequences. Perhaps, the best example of "site-specific" recombination is the integration of viral DNA into the host chromosome.

Transcription

What is transcription?

Transcription is the process by which genetic information is transferred from DNA to RNA. Thus, transcription is the biosynthesis of RNA molecules, each of which is complementary to only one of the two strands of a region of the DNA template.

What types of RNA transcripts are made in a cell?

Generally, three major types of transcripts are found in a cell. They include ribosomal RNA (rRNA), which is the most abundant of the types of RNA made on a mass basis, transfer RNA (tRNA), which is the next most abundant on a mass basis, and then messenger RNA (mRNA). There are also other species of small RNAs, and although their role in processing, editing, and affecting gene expression is important, they make up a very small percentage of total cellular RNA.

What is a gene?

A gene is generally defined as a region of DNA that is transcribed, and its boundaries are described by where transcription is initiated (I) and terminated (T). By convention, the DNA sequence of the nontemplate strand is used to discuss all base sequences of the gene and its regulatory regions (i.e., promoter). Importantly, the DNA sequence of the nontemplate strand within the boundaries of the gene will directly correspond to the sequence of the transcribed mRNA, except that uracils in the RNA sequence replace thymines in the DNA sequence. This convention has the advantage that the base sequence corresponding to amino acids (the codons) and start and stop signals for protein synthesis are in their standard forms and easily recognizable in the DNA sequence of the gene. Sequence 5' (or "upstream") of the transcription start site for the gene sequence is numbered with negative numbers. Sequence 5' (or "downstream") of the transcription start site for the gene sequence is numbered with positive numbers (Figure 4.1).

What are the general features of transcription?

Transcription is the absolutely asymmetrical meaning that only one strand of the DNA helix is copied during transcription of a gene. The synthesis of RNA is unidirectional, occurring in the 5'→3' direction. Each transcript has a unique length defined by a specific initiation and termination site.

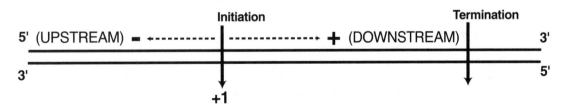

GENE

By convention, the sequence of the top strand (non-template strand), is used to discuss the base sequence of the gene.

Figure 4.1   Model of a gene. A gene may be considered a DNA segment that is transcribed. By convention the sequence of the top strand (the non-template strand) is used to discuss the base sequence of the gene and regulatory elements. The sequence of top strand that relates to the transcribed region will be identical in sequence to the transcribed mRNA, except that uracils in the RNA replace thymines in the DNA. The bottom strand is the template strand copied during transcription.

What are the common properties of all RNA polymerases?

RNA polymerases (DNA-dependent RNA polymerases) can self-initiate; that is, they do not require a primer to initiate. Synthesis is in the 5'→3' direction, and RNA polymerases are generally composed of multiple subunits.

Why is the *E. coli* RNA polymerase a good prototype for understanding RNA polymerase structure and function?

The *E. coli* RNA polymerase, like eukaryotic RNA polymerases, is composed on multiple subunits. The active form of the *E. coli* enzyme, holoenzyme, contains five different polypeptide chains: beta (β), beta prime (β'), two alpha (α) subunits, and sigma (σ). The holoenzyme has a molecular weight of approximately 450 kDa. The general structure, with two large subunits such as the β and β' and then multiple other subunits, is mimicked in eukaryotes. However, the eukaryotic RNA polymerases have more subunits and thus have a more complex structure.

Do the subunits of the *E. coli* RNA polymerase have known functions?

The β and β' subunits compose the actual transcriptional machinery, possessing the enzymatic activities required for both the initiation and transcriptional elongation. The much smaller α subunit has been assigned a role mainly in the assembly of the multisubunit complex, serving as a scaffold on which the rest of the complex is built. There is also evidence that α is involved in interaction with transcriptional regulators. The attachment of σ to the core enzyme is not very firm, and thus, it is relatively easy to isolate the core enzyme, which catalyzes the formation of internucleotide phosphodiester bonds equally well in the presence or absence of σ. However, the core enzyme does not display specific promoter recognition.

How does the eukaryotic transcription system differ from bacteria?

Transcription takes place in the nuclear compartment in eukaryotic cells where three classes of nuclear RNA polymerases exist. Unlike their bacterial counterpart, the eukaryotic RNA polymerases have essentially no intrinsic binding affinity for DNA.

What is the function of the three eukaryotic RNA polymerases?

The function of the eukaryotic RNA polymerases is specialized to the type of RNA synthesized. This was clearly demonstrated by their differential sensitivity to α-amanitin. The synthesis of ribosomal RNA by RNA polymerase I was found to be insensitive to α-amanitin inhibition. In contrast and at very low concentrations of α-amanitin, the synthesis of mRNA by RNA polymerase II was inhibited. At higher concentrations of α-amanitin, synthesis of tRNA and other small RNA species by RNA polymerase III was inhibited.

What are promoters?

The promoter is a regulatory region that directs the assembly of an initiation complex for transcription.

Are there common features to all eukaryotic promoters?

Sequence elements that contribute to promoter function are located in the correct 5'→3' orientation with transcription and on the correct strand of the double helix. These requirements support the proposal that promoter elements are required for the correct positioning of the RNA polymerase on the transcriptional start site.

Are there other regulatory elements that affect eukaryotic transcription?

In addition to the promoter, there are regulatory DNA elements that mediate an increase or a decrease in the transcription of neighboring genes, respectively, by enhancers and silencers. Positive control of transcription by enhancer elements and their binding proteins is more common than negative control by silencers. These regulatory elements (DNA sequences) are not constrained to the same strand of the helix or location as promoter elements. Although some enhancer elements are part of the promoter, others may be found several thousand base pairs upstream or downstream of the gene affected.

How do enhancers and silencer regulatory elements act?

Enhancers and silencers affect gene transcription by the binding of regulatory proteins that act as transcription factors, interacting to affect the activity of the transcriptional initiation complex. The sequence of the enhancer or silencer to which the regulatory protein binds is called a response element.

How does the interaction of transcriptional factors affect RNA polymerases?

As stated earlier, eukaryotic RNA polymerases do not possess the intrinsic ability to interact with DNA. Thus, promoter recognition is mediated by general transcription factors. General transcription factors are defined as DNA-binding proteins that are required for transcription of all genes by a particular RNA polymerase. In contrast, specific transcription factors are tissue specific and regulate transcription of some genes, but not all. Thus, the combination of general transcription factors with specific transcriptional factors is required for the RNA polymerase to initiate transcription from a specific promoter.

How is transcription regulated?

The ability to form an active transcriptional initiation complex efficiently is highly regulated by a large number of DNA-binding proteins. The efficiency of assembly of an active complex can be further regulated by additional regulatory elements to which proteins bind. These interact to facilitate or diminish the efficiency by which an active initiation complex forms.

Are there common features that contribute to the ability of proteins to interact with DNA?

Basic protein structure motifs have been identified that affect DNA binding and dimerization of transcription factors. For example, a DNA binding motif called a "helix-turn-helix" that involves two α-helices separated by a β-turn fits in the major groove of DNA. Zinc finger domains, which contain zinc bound by cysteine and histidine side chains, allow interaction through the major groove of DNA also. On the other hand, a leucine zipper motif represents two α-helices, one with basic residues for DNA binding and one with regularly spaced leucines for protein dimerization. The sequence elements of these regions are palindromes; that is, the sequence on the two strands of the helix reads the same.

How are transcription factors regulated?

Transcription factors are regulated at the level of their own synthesis, through interactions with other proteins and through covalent modification such as phosphorylation/dephosphorylation.

Does chromatin structure affect transcription?

The level of packaging of the DNA is very important to its accessibility for transcription. Therefore, the level of packaging and the interaction of histones with the DNA affect the ability to transcribe a gene. For this reason, DNA devoid of histones is easier to transcribe than nucleosome-associated DNA. Nucleosomes can be displaced from promoters and enhancers by the competitive binding of sequence specific transcription factors. In addition, modification of histone in nucleosomes can affect the degree to which interaction with DNA is maintained. For example, acetylation of histones in the nucleosome will diminish nucleosome structure and interaction with DNA, thus promoting transcription of that region of modified chromatin. On the other hand, deacetylation of the histones will enhance nucleosome structure and DNA packaging, thus hindering transcription of that region.

Does modification of the DNA template affect transcription?

Methylation of DNA at the 5-position of cytosine residues adjacent to guanosine residues in DNA (CpG regions) is present throughout the genome. Regions of DNA that are highly methylated are not transcribed. Therefore, the pattern and degree of DNA methylation can affect the genes that are transcribed in a particular region of the genome.

How do inhibitors of transcription work?

The functions of the RNA polymerase may be inhibited by compounds that interact directly with the DNA template or by compounds that bind directly to the RNA polymerase. For example, common inhibitors that bind to DNA directly and inhibit RNA polymerase are planar molecules with multiple aromatic rings. Typically, these types of inhibitors interact between the base pairs (intercalate) that span the double helix and have a high affinity for intercalation between G:C pairs. A common inhibitor of this type is actinomycin D in which the phenoxazone ring intercalates and the cyclic peptide portion hydrogen bonds specifically with guanosine residues above and below the phenoxazone ring in the minor groove. Actinomycin DNA preferentially blocks elongation, and different concentrations can have differential effects on RNA synthesis in the nucleus, nucleolus, and mitochondria. In contrast, only a few inhibitors of transcription are known to act specifically on RNA polymerases. These inhibitors have been important for differentiating and classifying RNA polymerases in prokaryotes and eukaryotes. For example, in prokaryotes, rifampicin (a rifamycin) binds specifically to the β-subunit of E. coli RNA polymerase, blocking incoming rNTPs and initiation, and streptolydigin inhibits E. coli RNA polymerase by blocking elongation. In eukaryotes, α-amanitin inhibits RNA polymerase II at a very low concentration and RNA polymerase III at a higher concentration.

5 Posttranscriptional RNA Processing

What is the role of posttranscriptional RNA processing in information metabolism?

> For most eukaryotic genes, posttranscriptional RNA processing is an essential step in the production of a functional RNA and serves as a versatile regulatory mechanism for the posttranscriptional control of gene expression.

How are alternative isoforms of a protein produced?

> Through the differential splicing of introns, two or more alternative isoforms of a protein can be produced from the same pre-mRNA transcript (Figure 5.1).

In addition to alternative splicing, are there other aspects of genomic structure that can generate multiple genes?

> Some genes may have multiple initiation sites for transcription, where upstream initiation may include an addition exon or expand the sequence for exon 1. Some genes may also have multiple polyadenylation sites, such that selection of a particular polyadenylation site may include or exclude potential exon(s). It is therefore important to point out that the unit length for most primary transcripts in human cells is determined by the site of initiation and by the polyadenylation site.

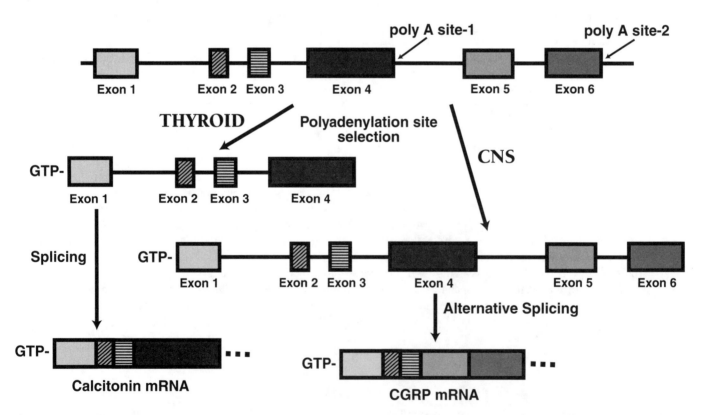

Figure 5.1 Differential gene expression involving posttranscriptional RNA processing. The differential expression of the calcitonin gene involves posttranscriptional processing where one of two different polyadenylation sites may be recognized. In the central nervous system, polyadenylation at the second site occurs along with alternative RNA splicing (exclusion of exon 4) to produce calcitonin-gene-related-peptide (CGRP). In thyroid tissue, the first polyadenylation site is recognized, excluding exon 5 and 6, followed by constitutive splicing of the first four exons.

How is pre-mRNA splicing regulated?

The process of pre-mRNA splicing is carried out by the spliceosome, where after the initial recognition of splice sites at the 3'-and 5'-ends of exons by splicing factors, two transesterification reactions take place, joining together the exons with the concomitant excision of the intron. The initial pairing of splice sites is a highly regulated process involving various *cis* and *trans* elements (Figure 5.2).

What are snRNPs?

These are small ribonucleoprotein particles (snRNPs), which represent a small nuclear RNA (snRNAs) complexed to proteins. In forming a spliceosome, snRNPs recognize and bind intro-exon splice sites by means of complementary sequences and interact with each other to facilitate the exchange of phosphodiester bonds (transesterification) that result in the removal of the intron and joining of exons.

What are the common *cis* elements on the pre-mRNA that affect splicing?

The *cis* elements on the pre-mRNA are the 5'-splice site and the 3'-splice site, which also include the branch point and a polypyrimidine tract.

What are pre-mRNA splicing enhancer elements?

Pre-mRNA enhancer elements are short RNA sequences capable of activating weak splice sites in nearby introns. These elements can activate heterologous pre-mRNAs and can function at a distance as great as 500 nucleotides from the affected intron.

How do pre-mRNA splicing enhancers work?

Similar to sequence elements that affect promoter function in transcriptional regulation, both constitutive and regulated splicing enhancers contain binding sites for SR proteins.

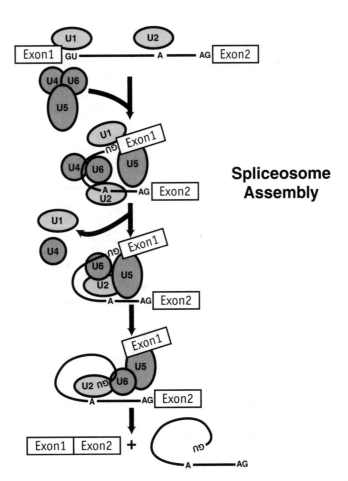

Figure 5.2 Proposed model for the mechanism of splicing. The various snRNPs are designated U1, U2, etc, and their orderly assembly results in the formation of a complex (spliceosome) with the pre-mRNA. Splicing involves a series of transesterification steps that result in the removal of the intron, and ligation of exon 1 to exon 2 in this model.

What are SR proteins?

SR proteins are a family of essential splicing factors containing one or more RNA recognition motifs (RNA binding sites) and an arginine/serine-rich domain. The RNA recognition motif is required for RNA binding, and the arginine/serine (RS) domain is required for protein–protein interactions. Mechanistic studies of splicing enhancer function are consistent with a recruitment model in which SR proteins activate splicing by binding to pre-mRNA splicing enhancer elements and recruiting the splicing machinery to the adjacent intron.

Is there evidence to suggest that transcription and RNA processing are coordinately regulated?

Although co-transcriptional RNA processing is not obligatory, depending on the gene, all three processing reactions (i.e., capping, splicing, and cleavage/polyadenylation) can be tightly coupled to RNA polymerase II transcription.

Does RNA splicing always require spliceosome machinery?

In the process of understanding RNA processing, it was noted that there were introns in pre-mRNA that did not require formation of a spliceosome. This type of intron was shown to be self-splicing where the intron itself possessed the necessary properties to catalyze its self-removal and splicing. Typically, these types of introns are found in lower eukaryotes, in mitochondrial pre-RNA, and in some bacteriophage mRNAs. The biological significant difference in self-splicing introns and those introns that rely on spliceosome mediated splicing is regulation. By regulating spliceosome assembly through splice site recognition, alternative patterns of splicing can be achieved.

What is RNA editing?

RNA editing is a posttranscriptional modification resulting in an alteration of the primary nucleotide sequence of RNA transcripts by a mechanism other than splicing. Editing may involve modification, insertion, deletion, or substitution of nucleotides in the RNA transcript. The substitution of one nucleotide for another has been observed in humans and can result in tissue specific differences in transcripts (Figure 5.3).

What are the potential roles of RNA editing by deamination in mammalian cells?

Edited RNA transcripts (A-to-I; C-to-U) possess sequences different from their unedited transcript counterparts and hence may display functional activities different from that shown by the unedited transcripts. Editing may alter processes including mRNA translation by changing codons and hence coding potential. Editing may alter pre-mRNA splicing patterns by changing splice site recognition sequences. Editing may affect RNA degradation by modifying RNA sequences involved in nuclease recognition. Editing may affect viral RNA genome stability by changing template and hence product sequences during RNA replication, and editing potentially may affect RNA structure-dependent activities that entail binding of RNA by proteins.

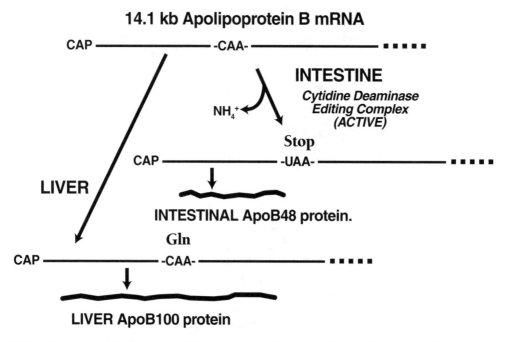

Figure 5.3 An RNA editing example. Tissue specific production of mammalian apolipoprotein (apo)B is regulated by RNA editing. This reaction involves deamination of a cytidine residue in a CAA codon to produce uridine and the in-frame UAA stop codon.

What editing appears to be the most widespread type of nuclear pre-mRNA editing in higher eukaryotes?

Adenosine to inosine modification in pre-mRNA, with inosine acting as guanosine during translation, is the most recent type of RNA editing to be discovered and appears to be the most widespread type of nuclear pre-mRNA editing in higher eukaryotes. For example, a well-studied editing event in humans is the modification of glutamate receptor subunit B pre-mRNA (GluR-B). Two edited sites (Q/R and R/G) are situated in the coding sequence of GluR-B. At the Q/R site, located in exon 11, a glutamine codon is changed to an arginine codon on editing. In the mammalian brain, Q/R is endogenously edited to nearly 100%. Receptors assembled with edited GluR-B subunits have reduced permeability for Ca^{2+} ions. The R/G site is located in exon 13, where the consequence of editing is a change of an encoded arginine to a glycine. The R/G site displays variable levels of editing during development. Receptors assembled with an R/G edited GluR-B subunit have been shown to recover faster from desensitization.

What is RNA interference?

RNA interference (RNAi) is an evolutionarily conserved process in which a double-stranded RNA (dsRNA) induces sequence-specific, gene silencing. The natural roles of RNAi have been suggested to include defense against viral infection and regulation of cellular gene expression.

How do small, non-coding dsRNAs induce posttranscriptional gene silencing?

The mechanism by which dsRNA induces gene silencing involves a two-step process. First, the dsRNA is recognized by the ribonuclease-III–like enzyme called Dicer, which cleaves a larger dsRNA into smaller dsRNAs of 21–23 nucleotides in length. These interfering RNAs are then incorporated into a multicomponent complex, which recognizes and targets a related sequence for either mRNA destruction or mRNA translational inhibition (Figure 5.4).

Can RNA interference transcriptionally silence gene expression?

Because RNA is able to base pair with DNA, the potential use of RNAi to target specific DNA sequences and silence transcription has been investigated. In yeast, small RNAs and the RNA silencing machinery have been shown to direct the formation of silent heterchromatin via methylation of histone H3. In another study, RNAi was targeted to CpG islands of a promoter of a human gene, resulting in silencing of expression through methylation of CpG sites, as well as methylation of histone H3.

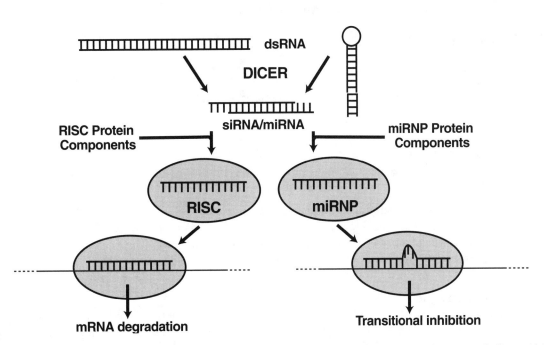

Figure 5.4   Proposed mechanisms for posttranscriptional targeted RNA interference. Long dsRNA and shRNA (short hairpin RNA) are processed to siRNA(short interfering RNA)/miRNA (micro RNA) duplexes by the RNase III-like Dicer. The short dsRNAs are subsequently unwound and assembled into effector complexes: RISC (RNA induced silencing complex) or miRNP (micro RNA nucleoprotein complex). RISC mediates mRNA-targeted degradation and miRNP guides translational inhibition of targeted mRNAs. In animals, siRNAs guide cleavage of complementary target mRNAs, whereas miRNAs mediate translational inhibition of mRNA targets.

6 Genetic Code and Protein Synthesis

How is genetic information decoded to direct the synthesis of protein?

The essential process involves the conversion of a 4-letter genetic alphabet to a 20-letter alphabet that describes a protein molecule.

What is the genetic code?

The genetic code is a nonoverlapping, degenerate code that makes use of a nucleotide triplet sequence, called a codon, to specify an amino acid or a stop signal.

What does the term *degenerate* mean relative to genetic code.

The term *degenerate* refers to the fact that multiple codons can specify the same amino acid. It is, however, important to note that no codon specifies more than one amino acid. For example, the amino acid arginine is specified by six different codons (Figure 6.1). There is no ambiguity in the use of codons.

How is the genetic code read?

The genetic code is read a codon at time during the process of protein synthesis. After the reading is initiated at a specific codon, there are no punctuations between codons and no overlap between codons. The information is read a codon at a time until a stop codon (nonsense codon) is encountered.

$$\underset{\text{Amino acid}}{\text{HCOO}-\overset{\overset{\displaystyle H}{|}}{\underset{\underset{\displaystyle NH_2}{|}}{C}}-R} + ATP + tRNA \quad \xrightarrow[\text{AMP + PPi}]{\substack{\textbf{\textit{Aminoacyl tRNA}} \\ \textbf{\textit{Synthetase}}}} \quad \underset{\text{AminoacyltRNA}}{tRNA-\underset{\underset{\displaystyle NH}{|}}{\overset{\overset{\displaystyle H}{|}}{OOC}-\overset{}{C}}-R}$$

What role does transfer RNA play in reading the genetic code?

Transfer RNAs are adapter molecules that translate the codons into the amino acid sequence of a protein. There is at least one species of transfer RNA (tRNA) for each of the 20 amino acids.

How does the right amino acid get connected to the correct tRNA?

The attachment of correct amino acid to the tRNA molecule is carried out in two steps by one enzyme for each of the 20 amino acids. This type of enzyme is called an aminoacyl-tRNA synthetase. The first step involves the activation of the amino acid forming an aminoacyl-AMP-enzyme complex. The second step recognizes the correct tRNA to which the aminoacyl moiety is attached through an ester linkage at the 3'-hydroxyl group of the terminal adenosine.

Do all tRNAs have common structural features?

The most common structure features of tRNA molecules are illustrated by the cloverleaf model for secondary structure. The based paired 5'-and 3'-ends represent the acceptor arm of the tRNA molecule, and the terminal nonbased paired 3'-CCA-OH end representing the attachment site for the amino acid. The thymidine-pseudouridine-cytidine arm is involved in binding of the aminoacyl-tRNA to the ribosomal surface, and the dihydrouracil arm (D arm) is apparently involved in aminoacyl-tRNA synthetase recognition. Finally, the anticodon loop of the cloverleaf model, the arm opposite the acceptor arm, recognizes the three-letter codon in mRNA (Figure 6.2).

Genetic Code

First Nucleotide	Second Nucleotide				Third Nucleotide
	U	C	A	G	
U	Phe	Ser	Tyr	Cys	U
	Phe	Ser	Tyr	Cys	C
	Leu	Ser	Stop	Stop	A
	Leu	Ser	Stop	Trp	G
C	Leu	Pro	His	Arg	U
	Leu	Pro	His	Arg	C
	Leu	Pro	Gln	Arg	A
	Leu	Pro	Gln	Arg	G
A	Ile	Thr	Asn	Ser	U
	Ile	Thr	Asn	Ser	C
	Ile	Thr	Lys	Arg	A
	Met	Thr	Lys	Arg	G
G	Val	Ala	Asp	Gly	U
	Val	Ala	Asp	Gly	C
	Val	Ala	Glu	Gly	A
	Val	Ala	Glu	Gly	G

Figure 6.1 Genetic Code. The codons are read from the table in the 5'→3' direction. All amino acids except methionine and tryptophan have more than one codon.

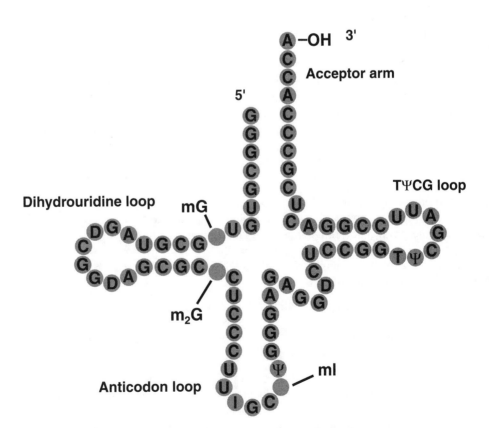

Figure 6.2 Cloverleaf model of yeast alanine tRNA (tRNA^{Ala}). Transfer RNAs base-pair with mRNA codons at a three-base sequence on the tRNA called the anticodon. The first base of the codon in mRNA (read in the 5'→3' direction) pairs with the third base of the anticodon. (modified residues: D= dihydrouridine, I = inosine, T = thymine, Ψ = pseudouridine, m = methyl group)

How does the anticodon recognize the correct codon in the messenger RNA?

Recognition involves, for the most part, appropriate base pairing (Watson-Crick base pairing). As with all base-paired interactions, the direction of the two base-paired strands run in opposite direction (antiparallel). Thus, the 5'-nucleotide of the codon in the mRNA is base paired with the 3'-nucleotide of the anticodon in the tRNA.

How can a tRNA recognize more than one codon specifying a particular amino acid?

The degeneracy of the genetic code resides mostly in the last nucleotide of the codon. As a result, there is some latitude in the base pairing that occurs between the 5'-nucleotide of the anticodon of the tRNA and the 3'-nucleotide of the codon of the mRNA. This is termed *wobble*.

Does codon recognition by a tRNA molecule depend on the amino acid that is attached?

Codon recognition does not depend on the amino acid attached to the tRNA. This was eloquently demonstrated by an experiment in which a cysteinyl-tRNA$_{cys}$ was chemically modified after charging such that the attached cysteine was converted to alanine. When this alanyl-tRNA$_{cys}$ was used in the in vitro translation of a hemoglobin mRNA, an alanine was incorporated at what was normally a cysteine site in the hemoglobin protein.

What is a ribosome?

The ribosome is the cellular component on which the various functional components interact to assemble the protein molecule. When many ribosomes assemble to translate simultaneously a single mRNA molecule, this is called a polyribosome.

Where does protein synthesis occur in the cell?

Generally, translation of messenger RNA occurs in the cytosol. The rough endoplasmic reticulum represents synthesis of proteins that are to be sorted for transport to other cellular compartments, become membrane associated, or are secreted. However, protein synthesis free of the endoplasmic reticulum may also include the synthesis of proteins that segregate to other compartments such as the peroxisomes, mitochondria, and nucleus, as well as for proteins that remain in the cytosolic compartment.

What is the structure of the ribosome?

Basically, the ribosome is composed of a large and a small subunit that were originally distinguished by their sedimentation constants. Although the size of the large and small subunits may vary between bacteria, lower eukaryotes, and higher eukaryotes, their role and essential aspects in protein synthesis are the same. For example, in higher eukaryotes, the large subunits are referred to as the 60S subunit, and the smaller is referred to as the 40S subunit. The 60S subunit is a ribonucleoprotein particle (RNP) because it is made of both RNA and protein components. The 60S ribosomal RNA (rRNA) components include a 28S rRNA, a 5.8S rRNA, a 5S rRNA, and as many as 50 different protein components. The 40S subunit is again a ribonucleoprotein particle made of an 18S rRNA and approximately 30 different protein molecules. In lower eukaryotes and bacteria, the sizes of their large and small subunits as well as the large and small rRNAs that make up these subunits are smaller.

How is the initiation codon selected?

Initiation of protein synthesis involves the formation of a preinitiation complex and includes tRNA, the small ribosomal subunit, and mRNA, with at least 10 eukaryotic initiation factors (eIFs). Also required are GTP and ATP. The formation of the preinitiation complex typically requires recognition of the 5'-cap structure (7-methylguanosine) and the binding of a ternary complex consisting of methionyl-tRNA$^i_{met}$, 40S ribosomal subunit, and eIF-2. After association of the 40S preinitiation complex with the mRNA cap, the complex scans the mRNA for a suitable initiation codon. Generally, this is the 5'-most AUG.

After the initiation codon is selected, how does translation proceed?

The binding of the 60S ribosomal subunit to the 40S preinitiation complex at an AUG start codon completes the formation of the 80S ribosome, resulting in the hydrolysis of GTP and the release of initiation factors. At this point, the methionyl-tRNA$^i_{met}$ is on the P site of the ribosome, ready for the elongation.

How does the correct aminoacyl-tRNA enter the A site on the ribosome-mRNA complex?

Elongation involves several steps catalyzed by proteins called elongation factors. These steps involve the binding of the correct aminoacyl-tRNA to the A site, peptide bond formation, and then translocation of the ribosome to the adjacent codon. The binding of the proper aminoacyl-tRNA in the A site requires proper codon recognition, an elongation factor that forms a complex with GTP and the entering aminoacyl-tRNA. This complex allows the aminoacyl-tRNA to enter the A site with the hydrolysis of GTP and the release of elongation factor-GDP complex.

How does a peptide bond form between the peptidyl-tRNA in the P site and the aminoacyl-tRNA in the A site?

The α-amino group of the new aminoacyl-tRNA occupying the A site carries out a nucleophilic attack on the esterified carboxyl group of the peptidyl-tRNA occupying the P site. This reaction is catalyzed by peptidyltransferase of the 60S ribosomal subunit and results in the attachment of the growing peptide to the tRNA in the A site.

After formation of the peptide bond, how does the ribosome move to the next codon?

After removal of the peptidyl group from the tRNA in the P site, the tRNA quickly dissociates. At this point, translocation occurs in a 5'→3' direction on the mRNA so that the peptidyl-tRNA at the A site moves into the empty P site on the ribosome, and the resulting empty A site is now in register with the next codon. This translocation requires elongation factor 2 (eEF-2) and GTP.

What terminates the translation process?

When the ribosome encounters a stop (nonsense) codon in the A site, after multiple cycles of elongation that polymerize the specific amino acids into a protein molecule, a releasing factor (eRF) capable of recognizing that termination codon in conjunction with GTP and peptidyltransferase promotes the hydrolysis of the bond between the peptide and the peptidyl-tRNA now occupying the P site. After hydrolysis and release of the peptide, the 80S ribosome dissociates into its 40S and 60S subunits, which can now be recycled for another round of translation.

Why are there antibiotics that only inhibit bacterial protein synthesis?

Ribosomes in bacteria (and in the mitochondria of eukaryotic cells) differ sufficiently from the mammalian ribosomes to be exploited by inhibitors that specifically interact with the bacteria translational machinery. The macrolide class of antibiotics such as tetracycline, lincomycin, erythromycin, and chloramphenicol do not interact with eukaryotic ribosomal particles but do bind to the 23S rRNA of the 30S subunit of bacterial ribosomes, inhibiting translation.

Are there antibiotics that inhibit eukaryotic protein synthesis?

Puromycin, an analogue of tyrosinyl-tRNA, effectively inhibits protein synthesis in both bacterial and eukaryotic systems. In contrast, cycloheximide inhibits the peptidyltransferase activity associated with the 60S subunit of eukaryotes only.

How does diphtheria toxin inhibit protein synthesis?

Corynebacterium diphtheriae, infected with a specific lysogenic phage, produces an exotoxin that catalyzes the ADP-ribosylation of eEF-2 in mammalian cells.

Do posttranscriptional modifications of pre-mRNA play a role in translation?

Evidence suggests that interactions between proteins that bind 7-methylguanosine cap and the poly(A) tail are involved in initiating the scan and assembly of a translation initiation complex at the initiator codon. Polyadenylate binding protein (PABP) and the poly(A) tail of eukaryotic mRNAs play an important role in stimulating translation initiation. PABP is comprised of two functional domains: an N-terminal domain with four RNA recognition motifs (RRM) and a carboxyl–terminus domain (CTD). eIF4G1 simultaneously binds the N-terminal domain of PABP (RRM2) and eIF4E in a way that facilitates the circularization of mRNA. This complex has been demonstrated to stimulate translation in mammalian systems. Although the mechanism is unclear, PABP binding has been proposed to induce cooperative conformational changes in eIF4E and eIF4G that enhance the stability of initiation complexes on capped mRNAs.

Do mRNAs have different levels of stability that affect their expression?

Although many studies tend to focus on transcription and pre-mRNA processing, cytoplasmic mRNA levels represent a balance between the rates of nuclear RNA synthesis, processing and export to the cytosol, and the rates of cytoplasmic mRNA degradation. Interestingly, there is an ever-increasing volume of information that has been accumulating, suggesting that mRNA decay can be a major control point in overall gene expression.

How is mRNA stability information conveyed?

Decisions as to the routing, translation, or disposal of an mRNA involves recognition of signals in coding and noncoding portions of each mRNA and the association of corresponding RNA-binding proteins.

What role does mRNA turnover play in gene expression?

Differential mRNA turnover may contribute to rapid changes in the pattern of cellular gene expression in response to changing environmental or developmental cues.

How does the turnover rate of an mRNA relate to the function of the encoded protein?

Messenger RNAs that encode proteins produced in short bursts (i.e., signaling proteins) in response to internal or external stimuli typically have short half-lives. These stimuli range from developmental (proliferating and differentiating), nutritional, hormonal, and pharmacologic to environmental alterations, such as temperature shifts and insults such as hypoxia and viral infection (Figure 6.3).

What features of the mRNA play a role in affecting mRNA turnover?

The regulated decay of mRNA is achieved by the coordinate interactions between an mRNA's structural components and specific *trans*-acting factors. These components include the 5'-cap structure, 5'-untranslated region (5'-UTR), the protein coding region, 3'-UTR, and the 3'-polyadenylate (polyA) tail.

Can a mutation in a gene affect the stability of an mRNA and contribute to disease state?

Stabilization of an otherwise unstable c-myc mRNA by reciprocal translocations of immunoglobulin introns in the 5'-UTRs of c-myc mRNA in certain lymphoma and plasmacytoma cells causes overexpression of the protein, which in turn promotes tumorigenesis.

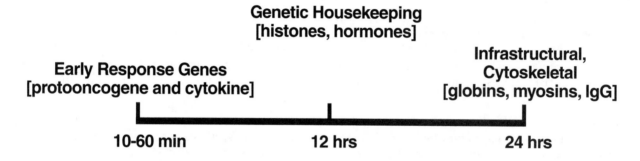

Figure 6.3   Relative stability of some mRNAs. The differential turnover of mRNA may contribute to rapid changes in the pattern of cellular gene expression in response to changing environmental or developmental cues, as well as to strictly maintain the levels of a translatable transcript under other situations.

Protein Structure and Function

What is the primary structure of a protein?

The primary structure of a protein is the order in which amino acids are joined together. Proteins are information containing molecules because the sequence or order of the amino acids is not random.

What are the unique properties of the peptide bond that affect secondary structure?

The peptide bond has a partial double-bond character that links the carbonyl carbon to the α-nitrogen of a peptide bond. This double-bond character significantly restricts possible conformations of its associated atoms, which become coplanar. Thus, rotation is allowed only about two of the three covalent bonds that form the polypeptide backbone of proteins.

What is the secondary structure of a protein?

This represents the conformation of the polypeptide backbones of proteins. Secondary structures include the α-helix, β-sheets, β-bends, and irregular conformations termed loops or coils.

What are the characteristics of an α-helix?

The α-helix has a hydrogen bonding pattern that confers maximum stability. There are 3.6 amino acid residues per turn of the helix and the aminoacyl R-groups are directed outward from the helix axis, minimizing steric interference. Because the α-helix is the lowest energy and most stable conformation for a polypeptide chain, it forms spontaneously. This low energy state primarily arises from the formation of the maximum number of hydrogen bonds between the amino nitrogen and the fourth carbonyl oxygen in line of the helix. In addition, the tightly packed atoms at the core of the α-helix are in van der Waals distances with one another across the axis of the helix.

What are van der Waals interactions?

These are extremely weak forces that act over very short distances. The van der Waals contact distance is the optimum distance at which an attractive force, induced by dipoles, is maximum and the repulsive force, caused when two atoms come so close that their electron orbitals overlap, is minimum.

Why does proline disrupt an α-helix?

The peptide nitrogen of a proline residue cannot form a hydrogen bond. Thus, proline can only fit in as the first amino acid of an α-helix; otherwise, proline produces a bend in the helix.

What does the term *amphipathic α-helix* mean?

This is a special case for an α-helix, in which residues switch between hydrophobic and hydrophilic every three or four residues of the helix such that one face of the helix is polar and the other is nonpolar.

How does β-pleated sheet secondary structure differ from the α-helix?

Rather than intrachain interactions, the β-pleated sheet involves interactions between adjacent polypeptide chains that may run in an antiparallel or parallel direction. Alternating pairs of narrowly and widely spaced hydrogen bonds stabilize the antiparallel or parallel sheets.

Are there other structures that contribute to the secondary structure of proteins?

Although half of the residues in a typical globular protein may be present in α-helix and β-pleated sheet secondary structure motifs, the remainder reside in "loop" or "coil" conformations that may be considered to be irregularly ordered but are still important to secondary structure considerations. It is also important to consider that these structures are not "random coil," which denotes disordered structure.

How does secondary structure affect tertiary structure?

The term *tertiary structure* refers to the spatial relationships between secondary structural elements. The secondary structures of large proteins are often organized as domains. Tertiary structure defines the ways in which protein folding can bring together amino acids far apart relative to primary structure considerations and the bonds that stabilize these conformations.

How are domains defined relative to large proteins?

Folding of the polypeptide within one domain usually occurs independently of folding of another domain within the same polypeptide chain. Domains often contribute discrete functions to the protein. An example would be where one domain of a protein specifies ligand binding, whereas another domain of the protein is required for interaction with another protein.

What types of interactions stabilize the tertiary structure of a protein?

Electrostatic interactions between oppositely charged R groups of amino acid residues help stabilize structure. Disulfide bonds, in which two cysteine residues are oxidized to form a disulfide covalent bridge (cystine), confer additional stability to the tertiary structure of a protein. Finally, hydrophobic interactions involving nonpolar side chains that associate in the interior of a protein also contribute significantly to the stability of protein structure. Because there may be many hydrophobic interactions within the interior of protein, these types of interactions often serve as the major contributor to maintaining protein tertiary structure (native structure).

What conditions are used to disrupt the native structure of proteins?

Considerations would involve temperature, as well as agents such as urea to disrupt hydrophobic interactions, and pH, which would disrupt ionic interactions. In cases in which tertiary structure is stabilized by disulfide bonds, the addition of a reducing agent, such as β-mercaptoethanol, would also be required. A loss of structure is called denaturation.

Fibrous Proteins

How do proteins that function in the structural organization of cells or in the extracellular matrix differ from globular proteins?

Often, atypical sequences of amino acids within the protein as well as specific secondary and tertiary structure distinguish fibrous from globular proteins.

Where are fibrous proteins found?

Fibrous proteins are important components in skin, tendon, bone, and muscle. Fibrous proteins in these tissues contribute to structure and motile functions.

Why is collagen most often used as a representative fibrous protein?

Collagen comprises about 25% of total protein in the human body. Therefore, collagen is important relative to its abundance, as well as an example of how primary, secondary, and tertiary structures contribute to its function as a structural protein.

Does collagen have an atypical primary sequence?

Collagen is made of three polypeptide chains referred to as α-chains. The sequence in the α-chain represents a repeat pattern of Gly-X-Pro/Hyp. Every third residue is a glycine. (It is important not to confuse the collagen α-chain term with α-helix, as there is no relationship.).

-**Gly**-Pro-Met-**Gly** Pro-Ser-**Gly**-Pro-Arg-**Gly**-Leu-Hyp-**Gly**-Pro-Hyp-**Gly**-Ala-Hyp-**Gly**-

Amino acid sequence of a section of the α1(I) chain of collagen.
(Every third residue is glycine.).

What is the importance of a glycine repeat at every third amino acid?

Glycine is the only amino acid lacking an R group. The nature of the three amino acid repeat in the presence of proline in the third position of the repeat causes the collagen α-chain to form a left-handed helical structure. In order for three α-chains to interact appropriately, the chains must cross over each other at glycine residues where a bulky side chain would not fit. The interaction of the three collagen α-chains forms a right-handed triple helix (Figure 7.1).

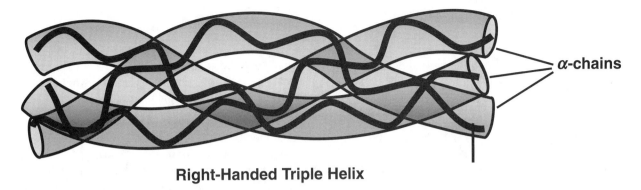

Right-Handed Triple Helix

α-chains

Figure 7.1 Collagen α-chains for a right-handed triple helix. The three α-chain helices wrap around one another as a right-handed triple helix. Glycine, because of its small size, is required at the tight junction where the three chains are in contact. The center of the triple helix is not hollow, but is very tightly packed when observed using a space filling model.

What is the significance of the prolyl residues?

The prolyl residues force the α-chain polypeptide into the extended left-handed helix. Hydroxylation of the prolyl groups, a posttranslational modification, contributes to the thermal stabilization of the right-handed triple helix formed from the interaction of three α-chains.

What is the source of the hydroxyproline and hydroxylysine in collagen?

Hydroxylation of proline and of some lysine residues is an important posttranslational modification of collagen catalyzed by specific enzymes called hydroxylases. Proline and lysine hydroxylases require ascorbic acid (vitamin C). Insufficient vitamin C results in a disease called scurvy, characterized by bleeding gums, poor wound healing, and ultimately death. Loss of this modification affects the stability of the collagen fibers and loss of function of collagen in tissue (Figure 7.2).

How are collagen fibers assembled?

Modification of side chains by hydroxylation and addition of a sugar group such as galactose stops after the three pro-α-chains interact. When the triple-helical molecule is packaged in secretion vesicles, specific proteases remove globular domains from the α-chains at both the carboxyl and amino terminal ends, which reduces the solubility and promotes interactions between the tropocollagen molecules. After release into the extracellular matrix, the association of tropocollagen molecules results in formation of collagen fibers. The structure of these fibers is further stabilized and strengthened as a result of covalent cross-links. Cross-linking results from the oxidation of the epsilon amino groups of lysine residues catalyzed by lysine oxidase, a copper-requiring enzyme. The resulting aldehyde generated on the R-side group of lysines reacts through an aldol condensation reaction.

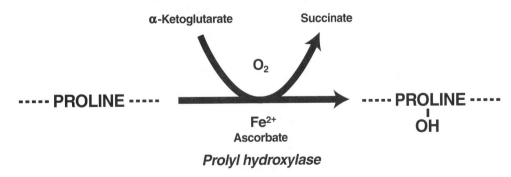

Figure 7.2 Reaction catalyzed by prolyl hydroxylase. The hydroxylated proline plays an essential role in the formation of collagen, and in maintaining structure. The collagen helix structure requires the proline residue in the third position of the Gly-X-Y repeat to assume a conformation that is enforced by the hydroxyl substitution at C-4 in 4-hydroxylproline. Thus, the inability to hydroxylate the proline at the third position of the repeat leads to collagen instability and connective tissue problems observed in scurvy.

Globular Proteins and Oxygen Carriers

What is myoglobin?

Myoglobin is a globular protein found in red muscle tissue that stores oxygen. Under conditions of oxygen deprivation, myoglobin releases oxygen for use by muscle mitochondria for oxidative phosphorylation. Myoglobin is composed of a single approximately 17-kDa polypeptide chain with Fe^{2+} iron and a heme moiety (Figure 7.3).

How does myoglobin differ from hemoglobin?

Hemoglobin is the major oxygen transport protein found in red blood cells. It is composed of four subunits, two α-globin subunits and two β-globin subunits, each of which represents an approximately 17-kDa polypeptide. Each globin subunit contains a heme and Fe^{2+} iron. Even though the globin chains are very similar in composition and structure to myoglobin, hemoglobin is composed of subunits and thus possesses quaternary structure that allows it to function as an efficient oxygen carrier in the blood stream.

Why is myoglobin unsuitable as an oxygen transport protein but effective for oxygen storage?

The partial pressure of oxygen in muscle tissue directly affects the amount of oxygen bound to myoglobin with hyperbolic oxygen-binding characteristics. As a result, myoglobin cannot give up oxygen readily until the partial pressure of oxygen is sufficiently low—for example, less than 20 mm Hg, observed in muscle tissue during active exercise. In contrast, hemoglobin will give up half of the oxygen carried per molecule at 26 mm Hg with sigmoidal oxygen-binding characteristics.

How does the binding of oxygen to hemoglobin vary with changes in the partial pressure of oxygen?

The oxygen binding curve for hemoglobin is sigmoidal. As a result, small changes in the partial pressure of oxygen will allow for significant levels of oxygen to bind or be released from hemoglobin (Figure 7.4).

What factors affect the binding of oxygen to hemoglobin?

Hemoglobin binds four oxygen molecules per tetrameric molecule. Because of the quaternary structure of hemoglobin, oxygen bound to one subunit can affect the binding of oxygen to another subunit. Thus, hemoglobin exhibits cooperative binding kinetics. Also, allosteric effectors can cause profound changes in oxygen binding. These factors include small changes in blood pH, oxygen tension, CO_2 concentrations in tissue, and the levels of 2,3-bisphosphoglycerate found in red blood cells.

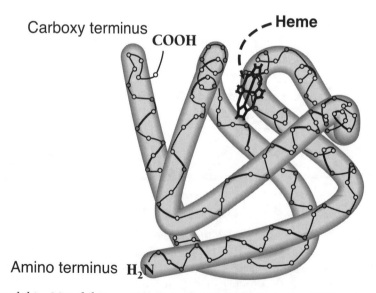

Figure 7.3 Structure of myoglobin. Myoglobin contains a single polypeptide chain of 153 amino acid residues, and a single iron protoporphyrin (heme) group. The backbone of the myoglobin molecule is made up of eight relatively straight segments of α-helix interrupted by bends.

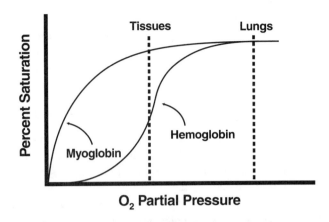

Figure 7.4 Oxygen binding curves for myoglobin and hemoglobin. The percent of oxygen-binding sites occupied is plotted against the concentration (partial pressure) of oxygen. The hyperbolic curve describes the binding of oxygen to myoglobin. The sigmoid binding curve represents hemoglobin oxygen binding demonstrating the cooperative binding effects of hemoglobin that render it more sensitive to small differences in oxygen concentration.

What is the Bohr effect?

This is the effect of pH hemoglobin oxygen binding. As the pH lowers in peripheral tissue, this lowering of pH shifts the sigmoidal curve for oxygen binding to the right. The result is a decreased affinity for oxygen at this lower pH. Thus, hemoglobin will give up bound oxygen more readily in tissue where the pH is lower. In the lungs, the effect is reversed. As oxygen binds to deoxygenated hemoglobin, protons are released and combine with bicarbonate, forming carbonic. Carbonic anhydrase in the red blood cells will aid in the production of CO_2.

$$CO_2 + H_2O \underset{Spontaneous}{\overset{Carbonic\ anhydrase}{\rightleftarrows}} H_2CO_3 \rightleftarrows HCO_3^- + H^+$$

Why does the pH of blood drop in peripheral tissues?

When CO_2, a product of metabolism, enters the blood from peripheral tissues, it forms carbonic acid, a weak acid that tends to lower the pH of the blood. In peripheral tissues, carbonic anhydrase in red blood cells will tend to enhance the formation of carbonic acid.

In addition to hemoglobin oxygen-carrying capacity, does hemoglobin also carry CO_2?

Hemoglobin facilitates the transport of CO_2 from tissues to the lungs for exhalation. Hemoglobin can bind CO_2 directly when oxygen is released. Hemoglobin carries approximately 15% of the CO_2 carried in blood. CO_2 reacts with the amino terminal α-amino acid of hemoglobin, forming a carbamate and releasing protons that contribute to the Bohr effect.

$$CO_2 + Hb\text{-}NH_3^+ \rightleftarrows Hb\text{-}NH\text{-}COO^- + 2H^+$$

How does 2,3-bisphosphoglycerate affect oxygen binding by hemoglobin?

One molecule of 2,3-bisphosphoglycerate (BPG) binds per hemoglobin tetramer in the central cavity formed by all four subunits. Binding of BPG is through salt bridges to the amino terminal ends of both β-chains. In this way, BPG stabilizes the deoxygenated form of hemoglobin and is the major effector that contributes to the cooperative binding of oxygen by hemoglobin.

Why does fetal hemoglobin ($\alpha_2\gamma_2$) have a greater affinity for oxygen than adult hemoglobin?

BPG binds more weakly to fetal hemoglobin (HbF) than to adult hemoglobin (HbA). As a result of an amino acid change in HbF that removes a salt bridge, BPG has a less profound effect on the stabilization of the deoxygenated form and is responsible for HbF appearing to have a higher affinity for oxygen than does HbA.

What is a hemoglobinapathy?

A hemoglobinapathy results from a mutation in hemoglobin that alters biologic function.

How does a hemoglobinapathy differ from a thalassemia?

In the case of a thalassemia, the mutation alters the production of hemoglobin as a result of either the absence or un-underproduction of a globin chain. For example, the altered production of the β-chain results in a β-thalassemia. When no β-chain is produced, this is referred to as β^0-thalassemia. When there is some β-chain produced, this is referred to as β^+-thalassemia. Depending on the levels of β-chain produced, patients may be asymptomatic to severely affected.

What is the molecular basis for sickle cell disease?

There is a single amino acid change, valine for glutamate, at the sixth position of the β-chain. This substitution replaces a polar glutamate residue with a nonpolar valine residue. As a result of this replacement, a "sticky patch" on the surface of the β-chain is generated. This hydrophobic patch provides a surface that complements another surface feature of hemoglobin present only in the deoxygenated form. As a result, HbS in the deoxyform polymerizes, forming long fibrous precipitates in the red blood cell. The red blood cell distorts, taking on a sickle shape, and is vulnerable to lysis as it penetrates the interstices of the splenic sinusoids.

What is hemoglobin A_{1C}?

Hemoglobin A_{1C} (HbA_{1c}) results from the nonenzymatic glycosylation of hemoglobin. The fraction of glycosylated hemoglobin is dependent on concentration of glucose in the blood. Approximately 5% of the hemoglobin is glycosylated in normal individuals. Elevation of this level is proportionate to blood glucose concentrations and thus provides information useful for the management of diabetic patients. Elevated HbA_{1C} reflects the average blood glucose concentration over a 6- to 8-week period, thus reflecting the long-term control of blood glucose levels.

Enzymes

General Properties

What are enzymes?

Enzymes are biological catalysts that participate in reactions in which substrates are converted to products. Enzymes therefore possess all of the properties that define a catalyst from chemistry, including properties such as the catalyst is neither formed nor consumed in a reaction and the catalyst affects the rate at which a reaction reaches equilibrium by lowering the energy of activation.

$$\text{Substrate} \overset{\textit{enzyme}}{\longleftrightarrow} \text{Product}$$

What biological molecules act as enzymes?

One major function of protein is to serve as enzymes. However, under special circumstances, RNA can act as an enzyme known as a ribozyme.

What is a coenzyme?

Some enzymes require a second organic molecule known as a coenzyme, without which they are inactive. Coenzymes are heat-stable, low molecular weight organic compounds required for enzyme activity.

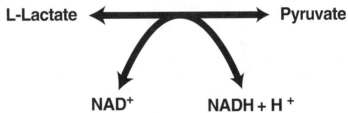

Lactic acid dehydrogenase

L-Lactate ⟷ Pyruvate

NAD⁺   NADH + H⁺

NAD acts as a coenzyme in the lactic acid dehydrogenase catalyzed reaction.

What properties of a reaction affect velocity?

The rate or velocity of a reaction is dependent on the amount of enzyme, the concentration of substrate available, the temperature, the pH, and even the ionic strength.

What type of curve is generated when the velocity (v) of an enzyme catalyzed reaction is plotted against the substrate concentration?

A hyperbolic curve is produced when velocity is plotted versus substrate concentration (Figure 8.1).

Can the rate of the reaction catalyzed by an enzyme no longer be dependent on substrate concentration?

At a given enzyme concentration, there are a defined number of active sites that at very high substrate concentration can be totally occupied (saturation) so that the rate of the reaction will no longer show dependence on the concentration of substrate. Therefore, the maximum rate for a given concentration of enzyme is achieved, and the rate is no longer responsive to the concentration of substrate. This rate is referred to as the maximum rate or velocity (V_{max}) for a given enzyme concentration.

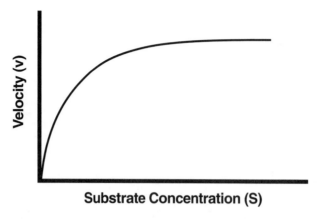

Figure 8.1 A plot of velocity versus substrate concentration. This plot demonstrates the effect of substrate concentration on the initial velocity of an enzyme-catalyzed reaction.

How does the Michaelis-Menten equation relate to an enzyme catalyzed reaction?

Michaelis-Menten proposed that during an enzyme catalyzed reaction, an enzyme–substrate complex is formed that may dissociate to reform the free enzyme and the substrate or react to release the product and regenerate free enzyme.

$$\text{Enzyme (E) + substrate (S)} \underset{k_2}{\overset{k_1}{\rightleftharpoons}} \text{Enzyme-substrate complex (ES)} \overset{k_3}{\rightarrow} \text{Enzyme (E) + Product (P)}.$$

where k_1, k_2, and k_3 are rate constants, and the Michaelis-Menten equations was derived:

$$v = \frac{V_m[S]}{K_m + [S]}$$

$$K_m = (k_2 + k_3)/k_1 \quad \text{and} \quad V_{max} = \text{is the maximum velocity}$$

What does the K_m tell us about the enzyme-catalyzed reaction?

When the substrate concentration equals K_m and is substituted into the Michaelis-Menten equation, velocity can be shown to equal $\frac{1}{2}V_{max}$. In other words, K_m is a kinetic term but is often important in describing an enzyme-catalyzed reaction because it provides a convenient way to define enzyme activity relative to substrate concentration. This becomes particularly important when one considers the activity of an enzyme in a cell where the concentration of substrate may be limiting.

Why is the Michaelis-Menton equation often transformed to the Lineweaver-Burk equation?

The Michaelis-Menton equation results in a plot that produces a hyperbolic curve that makes the graphical interpretation of V_{max} and K_m difficult. Conversion to the Lineweaver-Burk equation produces a plot with a straight line that allows an easier interpretation of K_m and V_{max} values (Figure 8.2).

$$1/v = K_m/V_m[S] + 1/V_m \quad \text{Lineweaver-Burk Equation}$$

How does the rate of catalyzed reaction vary with temperature?

The rate of a reaction increases with increased temperature, at least to the point until the enzyme may be inactivated through denaturation. The magnitude of this increase depends on the free energy of activation ΔG_{act}. The temperature dependence is called the Q_{10} value and represents the factor by which the reaction rate is increased by a temperature increase of 10°C. The greater the ΔG_{act} is, the greater the Q_{10} value.

Does the temperature dependence of a catalyzed reaction have clinical relevance?

The increased metabolic rate during fever is, at least in part, caused by the temperature dependence of enzyme-catalyzed reactions. It is also relates to the fact that the human body cannot tolerate temperatures greater than 42°C.

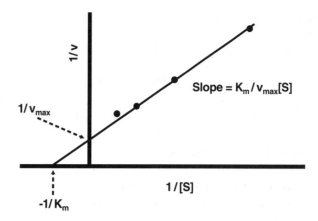

Figure 8.2 A double-reciprocal or Lineweaver-Burk plot. The double-reciprocal plot has the advantage of allowing an easier interpretation of V_{max} and K_m and distinguishing between certain types of enzymatic reaction mechanisms or analyzing enzyme inhibition. The line has a slope of K_m/V_{max}, an intercept of $1/V_{max}$ on the 1/V axis, and an intercept of $1/K_m$ on the 1/[S] axis.

At temperatures greater than this, some enzyme may denature. Most individuals can tolerate hypothermia far better than hyperthermia. Temperatures can decrease to close to 0°C where nerve conduction and muscular activity is blocked, but no irreversible damage will occur unless the water in tissue freezes.

Does the pH dependency of catalyzed reactions come into play in disease states?

Because ionizable groups in the active site of the enzyme whose protonation state depends on pH are often critical to substrate binding or catalysis itself, pH values in tissues and body fluids are tightly regulated. A change of more than 0.5 of a pH unit from the pH 7.4 for normal blood can be quickly fatal. Moreover, intracellular pH values are maintained within similarly strict ranges that depend on the cellular compartment. For example, the lysosomal pH ranges from 4.5 and 5.5, compared with a pH range of 6.5–7.0 for the cytoplasm.

What is the significance of the "rate limiting" catalyzed step in a metabolic pathway?

The significance of the catalyzed step in a metabolic pathway that is considered rate-limiting relates to consideration of the regulation of that pathway. In most instances, the step that is rate-limiting is also the committed step of the pathway. In other words, this step commits a particular metabolite to the end-product of that pathway. Most often, the rate-limiting or committed step is the step that is regulated. Regulation may involve the initial substrate of the pathway or feedback by the product. Hormones, through a signaling pathway, may upregulate or downregulate the activity of the enzyme catalyzing the rate-limiting step.

How does consideration of the rate-limiting step of a metabolic pathway relate to the clinical presentation of a patient?

A deficiency in the rate-limiting enzyme of a metabolic pathway will necessarily result in accumulation of the substrate for that enzyme and a deficiency of the product(s) of that pathway. Mutations that cause a deficiency may result in lower levels of the enzyme protein, lower specific activity of the enzyme itself, or loss of regulation. Clues in diagnosis would occur from the recognition of substrate accumulation in serum or urine and deficiency of the product of the pathway that may produce specific symptoms.

What is meant by allosteric regulation of an enzyme?

Enzymes under regulation of an allosteric effector show a sigmoidal plot of velocity versus substrate concentration rather than a hyperbolic plot. Typically, allosteric regulated enzymes are multisubunit proteins, and the allosteric effector may act positively by shifting the sigmoidal curve to the left or negatively by shifting the sigmoidal curve to the right. The physiologic importance of allosteric regulated enzymes is that a small change in substrate concentration may elicit a very profound change in velocity (Figure 8.3).

What are the effects of inhibitors on an enzyme-catalyzed reaction?

An inhibitor can directly compete against substrate for binding at the active site. Such an inhibitor is considered competitive and would affect the kinetics of the reaction by increasing the apparent K_m for the substrate. There would be no effect on the kinetic parameter V_{max} because a high enough substrate concentration can be achieved to compete out all affects of the competitive inhibitor. In contrast, a noncompetitive inhibitor does not compete with substrate for binding at the active site and reduces enzyme activity by binding at a site other than the substrate-binding site. Assuming that the binding of a noncompetitive inhibitor to the enzyme can occur with free enzyme or with the substrate–enzyme complex, the effect would be to lower V_{max} without affecting K_m. There are also inhibitors (uncompetitive) that bind only to the substrate-enzyme complex. This type of inhibition increases K_m and reduces V_{max}. Ultimately, there are inhibitors that irreversibly interact with an enzyme, and in this case, only V_{max} would be affected because the concentration of active enzyme would be directly proportional to the level of inhibitor. It is important to note that irreversible inhibition cannot be overcome by removal of the inhibitor but instead requires new synthesis of the enzyme.

Do concepts of enzyme inhibitors have relevance to clinical medicine?

Many therapeutic drugs are enzyme inhibitor based, and how the inhibitor represses a particular enzyme activity may be critical to its efficacy. Moreover, inhibitors may come in the form of toxic agents introduced from the environment or from infective agents. Consideration as to the nature of the inhibitor (drug or toxic agent) might be critical in the management of an affected patient.

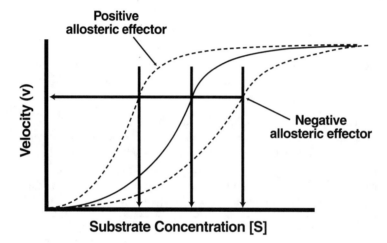

Figure 8.3 Plot of velocity versus substrate concentration for an allosterically regulated enzyme. Note the change in substrate concentration required to produce the same velocity for this enzyme by either a negative or positive allosteric regulator.

Carbohydrate Metabolism

Glycolysis

What is the significance of the first step in the commitment of glucose to glycolysis?

The first step in the commitment of glucose to glycolysis is the phosphorylation of glucose to glucose-6-phosphate, catalyzed by the enzyme hexokinase.

Is the hexokinase-catalyzed reaction specific for glycolysis?

hexokinase
$$glucose + ATP \rightarrow Glucose\text{-}6\text{-}phosphate + ADP$$

No. Glucose is always phosphorylated to glucose-6-phosphate before it is converted to other metabolic products. In most tissues (but not in liver), hexokinase is competitively inhibited by its own product, glucose-6-phosphate. This product inhibition prevents an accumulation of glucose-6-phosphate when the supply of glucose exceeds the capacity of the metabolizing pathways.

What is the metabolic significance of the phosphorylation of glucose by hexokinase?

This reaction commits glucose to intracellular metabolism because, unlike glucose, which can leave the cell on the same carrier on which it entered, glucose-6-phosphate is not transported across the plasma membrane.

How does glucokinase differ from hexokinase?

Glucokinase is not expressed in all tissues. In the liver, glucokinase is inducible, has a high K_m for glucose, and is specific for glucose.

What is the physiologic significance of glucokinase in the liver?

Because glucokinase is inducible and has a high K_m for glucose, it is poised to increase in activity as portal glucose increases above the normal 5 mmol/L (100 mg/dL). Unlike hexokinase, glucokinase is not inhibited by glucose-6-phosphate. Thus, the concentration of glucose-6-phosphate increases rapidly in liver after a carbohydrate-rich meal because of glucokinase activity.

Where is the glycolytic pathway located in cells?

The glycolytic pathway, so named because it is involved in the breakdown of carbohydrate (glucose), is a cytosolic pathway that is present in all cells.

What is the committed and rate-limiting step of glycolysis?

Phosphofructokinase-1
$$Fructose\text{-}6\text{-}phosphate + ATP \rightarrow Fructose\text{-}1,6\text{-}bisphosphate + ADP$$

In glycolysis, the conversion of glucose-6-phosphate to fructose-6-phosphate is a freely reversible reaction catalyzed by phosphohexose isomerase. However, when fructose-6-phosphate is then phosphorylated to fructose-1,6 bisphosphate by phosphofructokinase (PFK-1), this step is irreversible, committing the six-carbon intermediate, fructose-1,6-bisphosphate, to the glycolytic pathway.

What is the significance of triose isomerase step in glycolysis?

The six-carbon intermediate fructose-1,6-bisphosphate is cleaved into two triose phosphates by the enzyme aldolase where carbons 1, 2, and 3 form dihydroxyacetone phosphate, and carbons 4, 5, and 6 form glyceraldehyde-3-phosphate. These two triose phosphates are interconverted in a reversible reaction catalyzed by triose phosphate isomerase that ensures that all six glucose-derived carbons proceed through the glycolytic pathway.

Which is the first step in glycolysis in which energy is recovered?

Phosphoglycerate kinase
1,3-Bisphosphoglycerate + ADP → 3-Phosphoglycerate + ATP

The energy-rich intermediate 1,3-bisphosphoglycerate is converted to 3-phosphoglycerate by phosphoglycerate kinase where the phosphate group from the acyl-phosphate of 1,3-bisphosphoglycerate is transferred to ADP. This is a substrate level phosphorylation reaction, yielding the first ATP produced by glycolysis. Because the original six-carbon sugar was converted into two three-carbon compounds, two ATPs are actually derived from this step balancing the consumption of one ATP at the hexokinase step and one ATP at the phosphofructokinase step.

Why is the isomerization of 3-phosphoglycerate important in the overall production of energy?

The phosphate group in 3-phosphoglycerate is an ester phosphate and does not have sufficient energy to phosphorylate ADP. Through a series of isomerization reactions and then a dehydration reaction catalyzed by enolase, 3-phosphoglycerate is converted to a high-energy enol phosphate called phosphoenolpyruvate.

What step of glycolysis results in a net production of energy?

In the last reaction of glycolysis, catalyzed by pyruvate kinase, the high-energy phosphate group of phosphoenolpyruvate is transferred to ADP generating ATP. Again, because the original six-carbon sugar was converted into two three-carbon compounds, two ATPs are actually derived from this step on a net basis for each six-carbon unit that enters the glycolytic pathway (Figure 9.1).

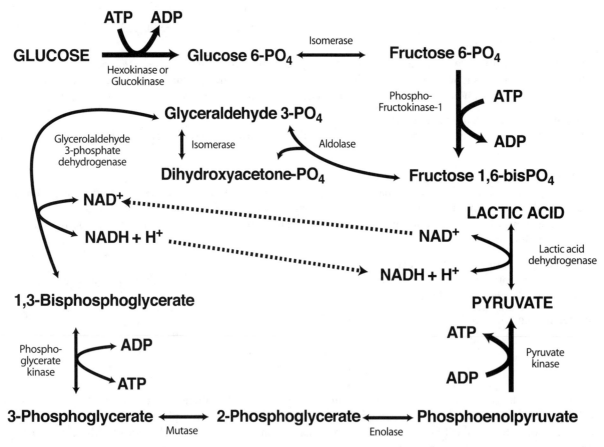

Figure 9.1   The glycolytic pathway. For each molecule of glucose that enters the pathway, two molecules of glyceraldehyde 3-phosphate are formed from fructose 1,6-bisphosphate.; thus, two molecules of pyruvate are produced as the end product. Under standard conditions in the cell, glycolysis is an essentially irreversible process, driven to completion by a large net decrease in free energy. Importantly, glycolysis releases only a small fraction of the total available energy of the glucose molecule. In order for glycolysis to continue under anaerobic conditions (i.e., exercising muscle tissue), the conversion of pyruvate to lactate is critical in order to regenerate oxidized NAD+ for the glycerolaldehyde 3-phosphate dehydrogenase step.

On a net basis, what products are generated by glycolysis?

Glycolysis generates two molecules of pyruvate, two molecules of ATP, and two molecules of NADH nicotinamide adenine dinucleotide (reduced) from each molecule of glucose that enters the glycolytic pathway.

What role does the lactate dehydrogenase (LDH)-catalyzed conversion of pyruvate play in maintaining glycolysis?

Glyceraldehyde 3-phosphate dehydrogenase (G3PHD)
Glyceraldehyde-3-phosphate + NAD⁺ + Pi → 1,3-Bisphosphoglycerate + NADH

Lactate dehydrogenase (LDH)
Pyruvate + NADH → Lactate + NAD⁺

Because NAD⁺ is present in only catalytic amounts in the cell, the reoxidation of the NADH, formed on reduction NAD in the G3PDH reaction, back to NAD⁺ is essential to allow glycolysis to continue. The reaction catalyzed by LDH is, therefore, particularly important in tissues that rely on glycolysis as their essential energy source—for example, muscle under anaerobic conditions and in red blood cells that lack mitochondria.

Can glucose be produced simply by reversing glycolysis?

No. Three steps are irreversible. The conversion of glucose to glucose-6-phosphate by hexokinase, and two steps in the glycolytic pathway catalyzed by phosphofructokinase and by pyruvate kinase are irreversible. All other glycolytic reactions can proceed in either direction under physiologic conditions.

Overall, how does the glycolytic pathway respond to fed and fasting conditions in most tissues?

It is important to note that the requirement for glycolytic activity varies in different physiologic states. Most tissues have increased glycolytic activity after a carbohydrate-rich meal, when excess dietary glucose has been metabolized. During fasting, most tissues reduce their glycolytic activity markedly while covering their energy needs by the oxidation of fatty acids and other alternative substrates.

What forms of controls are observed for glycolysis?

Regulation of glycolysis is dependent on the tissue and is far more complicated in liver, muscle, and other tissues than for red blood cells. The long-term regulation of glycolysis, particularly in the liver, is affected by changes in the levels of several glycolytic enzymes. These changes reflect alterations in the rates of enzyme synthesis and, possibly, degradation. For short-term regulation, the committed step catalyzed by PFK-1 is regulated. The regulation involves energy needs, such that ATP inhibits, whereas AMP and ADP stimulate this step. In the red blood cell, this step is regulated by energy levels of the cell such that the ratio of ATP to AMP and ADP, regulates the flux of glucose through glycolysis. In this same sense, the elevation of cytosolic citrate levels inhibits the activity of PFK-1. Citrate levels in nonred blood cells are commensurate with an "energy-rich" situation.

How do the hormones insulin and glucagon regulate PFK-1 activity in liver tissue?

In response to the fed situation, insulin levels increase and as a result stimulate liver PFK-1 in order to promote energy storage. In contrast, under fasting conditions, the elevation of serum glucagon levels signals inhibition of PFK-1 in the liver, and energy stores are mobilized to maintain serum glucose levels or another energy source.

Are there other steps in glycolysis that are regulated?

Pyruvate kinase is inhibited by phosphorylation in the liver (not muscle), and phosphorylation of pyruvate kinase is induced by glucagon under fasting conditions, enabling gluconeogenesis.

Why is anaerobic metabolism of glucose important in red blood cells?

The red blood cell is, both structurally and metabolically, the simplest cell in the body, losing all of its subcellular organelles during maturation. Therefore, the red blood cell relies exclusively on the anaerobic breakdown of glucose as its energy source.

What are the most significant energy demands of the red blood cell?

Most of the ATP (energy) used by red blood cells maintains the electrochemical and ion gradients across the plasma membrane. About 10% of glucose is diverted to the pentose phosphate pathway (hexose monophosphate shunt) to protect hemoglobin and other cellular constituents against oxidative damage. Another 10% to 20% is used for the synthesis of 2,3-bisphosphoglycerate, an allosteric regulator of the O_2 affinity of Hb.

How does the red blood cell synthesize 2,3-bisphosphoglycerate?

$$\text{1,3-Bisphosphoglycerate} \xrightarrow{\textit{Bisphosphoglycerate mutase}} \text{2,3-Bisphosphoglycerate} \xrightarrow{\textit{Bisphosphoglycerate mutase}} \text{3-Phosphoglycerate} + P_i$$

Bisphosphoglycerate mutase in the red blood cell converts 1,3-bisphosphoglycerate to 2,3 bisphosphoglycerate by-passing phosphoglycerate kinase catalyzed step. The bisphosphoglycerate mustase also has phosphatase activity that can convert 2, 3-bisphosphoglycerate to 3-phosphoglycerate.

Pentose Phosphate Pathway

Why is the pentose phosphate pathway often described as a shunt?

The pentose phosphate pathway is described as a shunt (hexose monophosphate shunt) because when pentoses are not needed for biosynthetic reactions, the pentose phosphate intermediates are cycled back to glycolysis as fructose-6-phosphate and glyceraldehyde-3-phosphate (Figure 9.2).

Where is the pentose phosphate pathway localized in the cell?

The pentose pathway is a cytosolic pathway present in all cells.

What is the essential function of the pentose phosphate pathway?

There are two major products of the pentose phosphate pathway: the pentose phosphates (ribose-5-phosphate) for synthesis of nucleotides and NADPH (nicotinamide adenine dinucleotide phosphate [reduced]). Again, depending on the tissue type, the pathway may be active for either of these two purposes. For example, in the red blood cell, a nondividing cell, there is limited need for nucleotide synthesis but an essential need for NADPH.

Why is there a significant requirement for the production of NADPH in the red blood cell?

NADPH is needed for the reduction of glutathione by glutathione reductase. Reduced glutathione is involved in important antioxidant systems in the red blood cell. It acts as a sulfhydryl buffer, maintaining -SH groups in proteins and enzymes in the reduced state, and it is used by glutathione peroxidase, a selenium containing enzyme, to reduce both H_2O_2 and organic peroxides.

Why has it been suggested that the pentose phosphate pathway has two parts, an oxidative and nonoxidative component?

The redox stage of the pentose phosphate pathway is involved in NADPH production, whereas the nonoxidative component of the pathway involves pentose phosphate production or interconversions that route the majority of the pentose phosphate back to glycolysis.

How is the pentose phosphate pathway regulated?

$$\text{glucose-6-phosphate} + NADP^+ \xrightarrow{\textit{glucose-6-phosphate dehydrogenase}} \text{6-phosphogluconate} + NADPH$$

Regulation occurs at the first committed step of the pathway catalyzed by glucose-6-phosphate dehydrogenase from the oxidative side of the pathway. NADPH feedbacks and inhibits glucose-6-phosphate dehydrogenase as a negative allosteric effector. In the red blood cell, under conditions of oxidative stress, such as drug-induced hemolytic anemia, the demand for NADPH is greatly increased. In response, fructose-6-phosphate produced by the nonoxidative side of the pathway can be converted to glucose-6-phosphate by phosphoglucose isomerase reaction and recycled through the pentose pathway to enhance production.

Where in the pentose phosphate pathway is ribose-5-phosphate produced?

Ribulose-5-phosphate can subsequently be converted to ribose-5-phosphate by an epimerase. When demands for nucleotide synthesis are lessened, both ribulose-5-phosphate and ribose-5-phosphate are converted back to two fructose-6-phosphates and one glyceraldehyde-3-phosphate.

Why is thiamine (vitamin B_1) required for the pentose phosphate pathway?

Transketolase contains a tightly bound thiamine pyrophosphate as its prosthetic group and catalyzes the transfer of a C_2 unit designated as an active "glycolaldehyde" from xylulose-5-phosphate to ribose-5-phosphate to produce the C_7 sugar sedoheptulose and glyceraldehyde-3-phosphate in the pentose phosphate pathway.

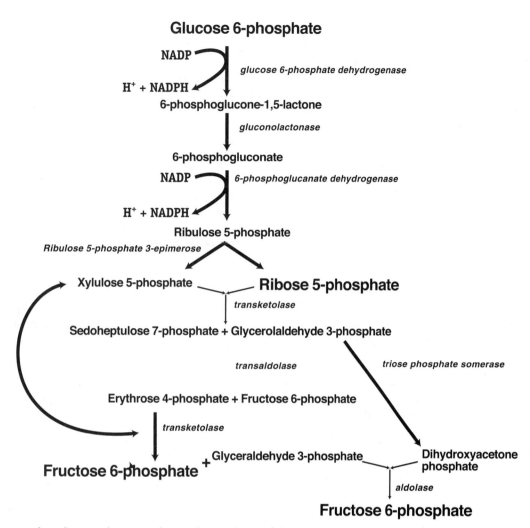

Figure 9.2 Pentose phosphate pathway. In the oxidative phase of the pentose phosphate pathway, NADPH is formed, which serves to support reductive biosynthesis. The other product of the oxidative phase is ribose-5-phosphate, which serves as a precursor for nucleotide and coenzyme biosynthesis.

In addition to its requirement to maintain antioxidant defenses, what other role does NADPH serve?

In tissue, other than the red blood cell, NADPH produced by the pentose phosphate pathway is used by synthetic (anabolic) pathways. As an example, the NADPH produced by the oxidative side of the pentose phosphate pathway is used in fatty acid synthesis.

Why do some individuals with a genetic defect in glucose-6-phosphate dehydrogenase present with jaundice and anemia when treated with anti-malaria drugs?

A number of drugs, particularly primaquine and related antimalarials, undergo redox reactions in the cell, producing large quantities of superoxide and H_2O_2. Superoxide dismutase converts superoxide into H_2O_2, which is inactivated by glutathione peroxidase, using NADPH as coenzyme. Because of insufficient production of NADPH by the mutant enzyme under stress, the red blood cell's ability to recycle oxidized glutathione (GSSG) to reduced glutathione (GSH) is impaired, and drug-induced oxidative stress leads to lysis (hemolysis) and hemolytic anemia.

Which tissues of the body absolutely require blood glucose for energy metabolism?

Red blood cells and the brain consume about 80% of the 200 g of glucose consumed in the body per day and have an absolute requirement for blood glucose for energy metabolism. Because the maximum level of blood glucose in the total plasma and extracellular fluid volume is approximately 10 g, blood glucose must be replenished constantly.

Glycogen Metabolism

What polysaccharide is a storage form of glucose?

During and immediately after a meal, glucose is converted in liver into the storage polysaccharide glycogen by a process known as glycogenesis. Between meals, glycogen is gradually degraded by the pathway of glycogenolysis. Ultimately, in the liver, the glucose derived from glycogen breakdown is released to the blood to maintain the glucose levels. Although glycogen is also stored in muscle, this glycogen is not available to maintain blood glucose.

What is glycogen?

Glycogen is a branched polysaccharide storage for glucose that contains two types of glycosidic linkages, extended chains of $\alpha1\rightarrow4$ linked glucose residues with $\alpha1\rightarrow6$ branches spaced about every four to six residues along the $\alpha1\rightarrow4$ chain. The gross structure of glycogen is dendritic in nature, expanding from a core sequence bound to a protein (glycogenin) through a tyrosine residue (Figure 9.3).

Where is glycogen stored?

Glycogen is stored in two tissues. In the liver, glycogen is stored for the short-termed maintenance of blood glucose. In muscle, glycogen is stored as a source of energy.

What step is regulated in the synthesis of glycogen?

Using UDP-glucose (activated sugar), glycogen synthase catalyzes the transfer of glucose to glycogen through an $\alpha1\rightarrow4$ linkage. This is the regulated step in glycogen synthesis (glycogenesis). The unphosphorylated form of glycogen synthase is active (high insulin, fed), whereas the phosphorylated form of glycogen synthase is inactive (high glucagon, fasting).

Could regulation of glycogen occur at the synthesis of UDP-glucose?

No, UDP-glucose pyrophosphorylase is not regulated because UDP-glucose is also used for the synthesis of glycoproteins, glycolipids, and other sugars.

How is the branched structure of glycogen maintained?

When the formation of $\alpha1\rightarrow4$ linkages exceeds eight residues in length, glycogen branching enzyme, a transglycosylase, transfers some of the $\alpha1\rightarrow4$ linked sugars to an $\alpha1\rightarrow6$ branch, setting the stage for continued elongation of both $\alpha1\rightarrow4$ chains until they, in turn, become long enough for transfer by branching enzyme.

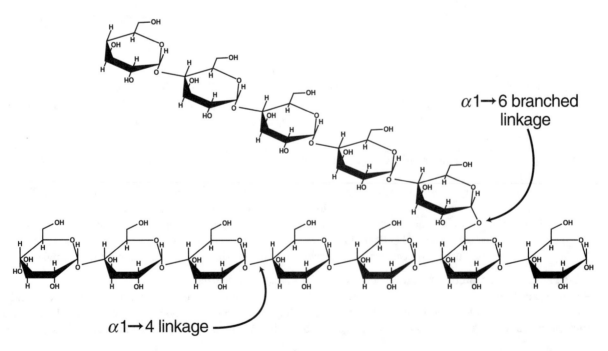

Figure 9.3 Glycogen branched chain. In vertebrates, glycogen is found primarily in the liver and skeletal muscle. Glycogen granules are complex aggregates of glycogen and the enzymes that synthesize and degrade it.

How is glycogen synthesis (glycogenesis) regulated?

After a meal, insulin is secreted to affect the efficient use and storage of blood glucose. Insulin activates glycogen synthase and turns off glycogen phosphorylase, thus promoting glucose storage. Insulin also stimulates glucose uptake in muscle and adipose tissue.

How is glycogen degraded (glycogenolysis)?

The pathway of glycogenolysis begins with removal of the external $\alpha 1 \rightarrow 4$ linked glucose residues in glycogen by glycogen phosphorylase, an enzyme that uses cytosolic phosphate and releases glucose from glycogen in the form of glucose-1-phosphate. This is the rate-limiting and regulated step of glycogenolysis.

$$\text{Glycogen}_{(n)} + Pi \xrightarrow[phosphorylase]{Glycogen} \text{Glycogen}_{(n-1)} + \text{glucose-1-phosphate}$$

Why does glycogen degradation require other activities in addition to glycogen phosphorylase?

Glycogen phosphorylase is specific for $\alpha 1 \rightarrow 4$ glycosidic linkages. It cannot cleave $\alpha 1 \rightarrow 6$ linkages, and it cannot cleave linkages up to the branching glucose residues efficiently. Because of this, glycogen phosphorylase cleaves the external glucose residues until the branch is three to four residues long. Then, the debranching enzyme, which has both transglycosylase and glucosidase activity, moves this short branch attached to the $\alpha 1 \rightarrow 6$ branch to the end of an adjacent $\alpha 1 \rightarrow 4$ chain, leaving a single glucose residue at the branch point. The exo-1,6-glucosidase activity of the debranching enzyme removes this single glucose residue as free glucose. Overall, approximately 90% of the glucose released from glycogen is in the form of glucose-1-phosphate. The remainder is free glucose that comes from each of the $\alpha 1 \rightarrow 6$ branch points.

Under what conditions is glycogen breakdown (glycogenolysis) activated?

Glycogenolysis is active in response to both acute and chronic stress. This stress may be manifested by fasting conditions, in response to increased blood glucose use during prolonged exercise, in response to blood loss, and in response to acute or chronic threats.

How is glycogen breakdown regulated?

Glycogenolysis in the liver is activated in response to increased demand for blood glucose. There are three major hormonal activators: glucagon, epinephrine (adrenalin), and cortisol. Glucagon typically mediates a response to fasting conditions, where there is a need to maintain blood glucose levels. Epinephrine and glucagon mediate the response during acute stress and prolonged exercise. Cortisol may also induce a response caused by both psychologic and environmental stresses.

How does glucagon activate glycogen phosphorylase?

Glucagon binds to a hepatic-plasma membrane receptor that initiates (via cAMP-dependent protein kinase, protein kinase A) a cascade of reactions that leads to the phosphorylation of glycogen phosphorylase.

$$\text{Glycogen phosphorylase} + ATP \xrightarrow{Phosphorylase\ kinase} \text{Glycogen phosphorylase-Pi}$$
$$\text{(Inactive)} \qquad\qquad\qquad \text{(Active)}$$

What role do G proteins play in signal transduction by glucagon?

G proteins are trimeric, plasma-membrane, guanosine-nucleotide binding proteins that are involved in signal transduction for a wide variety of hormones. Binding of glucagon to its plasma membrane receptor stimulates exchange of GDP for GTP on the G protein. In turn, the G protein with bound GTP undergoes a conformational change that leads to dissociation of one of its subunits, which binds to and activates the plasma membrane enzyme, adenylate cyclase. The activated adenylate cyclase converts ATP to cyclic-3',5'-AMP (cAMP) that binds to and thus activates protein kinase A. Protein kinase A catalyzes the phosphorylation of phosphorylase kinase, which in turn activates glycogen phosphorylase through phosphorylation (Figure 9.4).

$$\text{ATP} \xrightarrow{Adenylate\ cyclase} \text{cAMP} + PPi$$

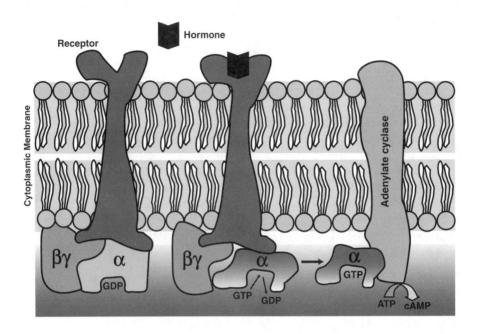

Figure 9.4 G protein activation of adenylate cyclase. Transduction of a signal through the β-adrenergic pathway involves seven steps that couples hormone binding to the receptor with activation of adenylate cyclase. For example, the binding of epinephrine to its receptor promotes a conformational change in the receptor's intracellular domain that affects its interaction with the second protein in the signal-transduction pathway, a heterotrimeric ($\alpha\beta\gamma$) GTP-binding stimulatory G protein, on the cytosolic side of the plasma membrane. When GTP is bound, this stimulates production of cAMP by adenylate cyclase.

Are there processes that ensure that this signaling is transient?

Phosphodiesterase activities that are also regulated, act on cAMP, hydrolyzing cAMP to produce 5'-AMP.

Phosphodiesterase
cAMP $\rightarrow$ 5'-AMP

Is the activation, via phosphorylation, of glycogen phosphorylase reversible?

Protein phosphatase catalyzes the removal of phosphate groups from glycogen phosphorylase, which in turn results in the inactivation of enzyme activity.

Protein phosphatase
Glycogen phosphorylase-Pi $\rightarrow$ Glycogen phosphorylase + Pi
(Active)　　　　　　　　　　　(Inactive)

Does epinephrine directly affect hepatic glycogen stores?

In liver, epinephrine mediates a response through several different types of receptors. During severe hypoglycemia, glucagon and epinephrine act to enhance glycogenolysis, where epinephrine acts through α- and β-adrenergic receptors. When blood glucose levels are normal, epinephrine can still mediate a response to increase blood glucose levels through the "fight or flight" response.

How does epinephrine affect glycogenolysis in liver?

Epinephrine mediates a response through β-adrenergic receptors via G proteins and cAMP the same way as glucagon. However, epinephrine also works through α-adrenergic receptors simultaneously, again through G proteins, but this time it involves activation of membrane isozyme of phospholipase C, which acts on phosphoinositol bisphosphate, releasing diacylglycerol and inositol trisphosphate as second messengers of epinephrine action. Diacylglycerol in turn activates protein kinase C, which initiates a series of protein–phosphorylation reactions. Inositol triphosphate promotes Ca^{2+} transport into the cytosol that then binds cytoplasmic calmodulin, which in turn activates phosphorylase kinase. The result is the activation of glycogen phosphorylase and increased glycogenolysis.

How does caffeine affect the epinephrine response?

Caffeine is a phosphodiesterase inhibitor, thus causing an increase in cAMP by inhibiting its breakdown. Consequently, there is an increase in blood glucose.

Can a deficiency arise in the ability of a tissue to break down glycogen?

Grouped under glycogen storage diseases, deficiencies in critical enzymes involved in glycogen metabolism can prevent the use of glycogen as a source of glucose. For example, a deficiency in debranching enzyme (Cori's disease or type III glycogen storage disease) leads to hypoglycemia and accumulation of glycogen stores in the liver. McArdle's disease or type V glycogen storage disease results from a deficiency in muscle phosphorylase. In this case, excess glycogen deposition is observed in muscle tissue, and patients with this disease suffer fatigue and muscle cramps induced by exercise. More common is type I glycogen storage disease (Von Gierke's disease), which results from a deficiency in liver glucose-6-phosphatase. This prevents dephosphorylation of glucose in the liver and the appropriate maintenance of blood glucose.

Gluconeogenesis

In addition to glycogen breakdown, what other metabolic process helps to maintain blood glucose levels?

The liver can also use amino acids from muscle protein as the primary precursor of glucose, as well as lactate from glycolysis, and glycerol from fat catabolism to synthesize glucose. This process, called gluconeogenesis, is essential for the maintenance of blood glucose during fasting or starvation.

What is the principle purpose of gluconeogenesis?

During fasting and starvation, when hepatic glycogen is depleted, gluconeogenesis is essential for the maintenance of blood glucose levels.

Where do the carbon skeletons for glucose production (gluconeogenesis) come from?

Three sources provide the carbon backbone for glucose synthesis during fasting: Lactate that is produced from anaerobic glycolysis in red blood cells and muscle, amino acids derived from muscle, and glycerol released from triglycerides during lipolysis in adipose tissue.

Of these sources, which is the major contributor to maintaining blood glucose levels?

Muscle protein is the major precursor of blood glucose during fasting and starvation. During prolonged fasting and starvation, both adipose mass and muscle mass are lost. The fat is used both for general energy needs and to support gluconeogenesis, whereas most of the amino acids from protein degradation are converted into glucose.

What is gluconeogenesis?

Conceptually, gluconeogenesis is the opposite of anaerobic glycolysis. From lactate, glucose is derived by partially using some glycolytic enzymes.

Why is gluconeogenesis not the reverse of anaerobic glycolysis?

There are three irreversible kinase reactions of the glycolytic pathway that must be overcome in order to generate glucose from pyruvate or lactate. These are glucokinase, phosphofructokinase-1, and pyruvate kinase.

How are the irreversible kinase reactions of the glycolytic pathway overcome to allow for gluconeogenesis?

Lactate is converted to pyruvate by lactic acid dehydrogenase, which then enters the mitochondrion. In the mitochondrion, pyruvate is converted to oxaloacetate by pyruvate carboxylase using biotin and ATP. Oxaloacetate is then converted to malate so that the malate can move out of the mitochondrion to the cytosol where it is reconverted back to oxaloacetate by cytosolic malate dehydrogenase. Oxaloacetate is then decarboxylated using GTP to yield phosphoenolpyruvate. Using the reverse direction of glycolytic enzymes, phosphoenolpyruvate can proceed back to fructose-1,6-bisphosphate. In order to bypass phosphofructokinase-1, another enzyme, fructose-6-phosphatase, converts fructose-1,6-bisphosphate to fructose-6-phosphate. Conversion of fructose-6-phosphate to glucose-6-phosphate occurs readily so that the last reaction to bypass glucokinase involves the enzyme glucose-6-phosphatase. This enzyme is associated with the endoplasmic reticulum and catalyzes the translocation and dephosphorylation of glucose to the lumen of the endoplasmic reticulum for the transport of glucose to the blood.

Why does a deficiency in glucose-6-phosphatase (type I glycogen storage disease, von Gierke's disease) result in serum lipid abnormalities such as increases in triglyceride, cholesterol, and phospholipids?

The metabolic state in the liver that results from glucose-6-phosphatase deficiency is essentially the same as that in diabetic ketoacidosis. The build up of glucose-6-phosphate in liver tissue results in an increase in glucose-derived acetyl-CoA. Elevated acetyl-CoA leads to activation of the lipogenic regulatory enzyme, acetyl-CoA carboxylase, which subsequently results in elevated malonyl Co-A. As a consequence, fatty acid oxidation is inhibited, and there is an increased synthesis and release of triglycerides from the liver.

Where does the energy come from to drive gluconeogenesis?

Although gluconeogenesis in the liver is efficient, it is also energy expensive, requiring a net expenditure of 4 moles of ATP per mole pyruvate converted to glucose. This ATP is provided by the oxidation of fatty acids.

How do amino acids get converted to glucose?

The backbone of amino acids can be converted either to glucose or acetyl-CoA or to both. After the removal of the amine group, the keto acid form of the amino acid may enter as an intermediate of the TCA cycle, via pyruvate or as an intermediate of the glycolytic pathway.

How is glucose derived from glycerol?

In liver, glycerol is phosphorylated by glycerol kinase and then enters gluconeogenesis pathway via the action of glycerol-3-phosphate dehydrogenase as dihydroxyacetone phosphate. Adipose tissue does not express glycerol kinase and therefore cannot use glycerol directly to generate glucose.

What is the source of glycerol for gluconeogenesis in the liver?

When fat stores are mobilized for energy as a result of lipolysis, glycerol is released from the hydrolysis of triglycerides. In particular, glycerol is released from adipose tissue to be taken up by the liver and converted to glucose during active gluconeogenesis.

Can the acetyl-CoA derived from fatty acid oxidation be converted to glucose?

Acetyl-CoA cannot be directly converted to pyruvate on a net basis, since the two-carbon backbone of acetyl-CoA cannot be converted to oxaloacetate via the TCA cycle because two carbons are lost as CO_2.

How is gluconeogenesis regulated?

Like glycogen metabolism, gluconeogenesis is primarily regulated by hormones. The rate-limiting step in gluconeogenesis is phosphoenolpyruvate carboxykinase, and this activity is upregulated by glucagon. In addition, the liver pyruvate kinase isoform is phosphorylated to an inactive form to prevent a futile cycle and the conversion of phosphoenolpyruvate back to pyruvate. Another futile cycle is prevented by regulation of phosfructokinase-1 and fructose-1,6-bisphosphatase. In this case, the allosteric regulator, fructose-2,6-bisphosphate, an activator of glycolysis, is reduced by the action of glucagon by phosphorylation of the bifunctional enzyme, phosphofructokinase-2/fructose-2,6-bisphosphatase. Phosphorylation inhibits the kinase activity so that the bifunctional enzyme acts as a phosphatase, reducing the levels of fructose-2,6-biphosphate and consequently PFK-1 activity. This result also relieves inhibition of fructose-1,6-bisphosphatase promoting gluconeogenesis (Figure 9.5).

How does increased degradation of fatty acids promote gluconeogenesis?

Oxidation of fatty acids produces acetyl-CoA, which enters the mitochondrion. Elevated acetyl-CoA levels inhibit pyruvate dehydrogenase (PDH), and acetyl-CoA is an essential allosteric activator of pyruvate carboxylase. Thus, pyruvate oxidation is inhibited and the conversion of pyruvate to oxaloacetate is favored, supporting gluconeogenesis.

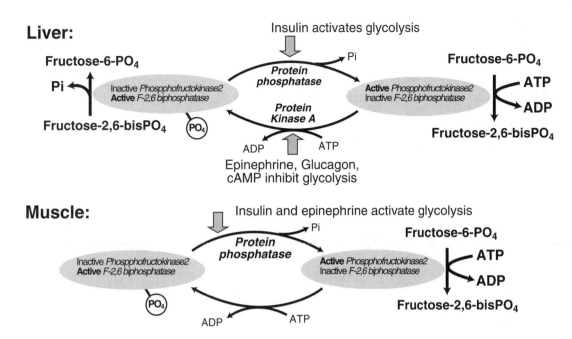

Figure 9.5 Regulation of fructose-2,6-biphosphate production. The cellular concentration of the regulator fructose-2, 6-bisphosphate is determined by the rates of its synthesis by phosphofructokinase-2 and breakdown by fructose-2, 6-bisphosphatase, which reside on the same polypeptide.

CHAPTER

10 | TCA Cycle and Oxidative Phosphorylation

Pyruvate Dehydrogenase

Where is pyruvate dehydrogenase located in the cell?

Pyruvate dehydrogenase is located in the mitochondrion.

What role does pyruvate dehydrogenase (PDH) play in TCA cycle function?

PDH catalyzes the conversion of pyruvate to acetyl-CoA. Thus, PDH acts to control the flux of acetyl-CoA available for the TCA cycle.

$$\textit{Pyruvate} \atop \textit{dehydrogenase}$$
$$\text{Pyruvate} + NAD^+ \rightarrow \text{Acetyl-CoA} + CO_2 + NADH + H^+$$

How is pyruvate dehydrogenase regulated?

Pyruvate dehydrogenase complex has associated kinase and phosphatase activities. The kinase activity is activated by high acetyl-CoA and NADH, which inactivates pyruvate dehydrogenase by phosphorylation. Pyruvate is then diverted to pyruvate carboxylase and oxaloacetate synthesis, which is important in gluconeogenesis. In contrast, pyruvate, CoA, and NAD^+ activate the phosphatase activity. Dephosphorylation of pyruvate dehydrogenase activates the enzyme; thus, pyruvate is now directed to acetyl-CoA production.

What are the unique features of pyruvate dehydrogenase?

Pyruvate dehydrogenase is structurally a multienzyme complex similar to several other α-keto acid dehydrogenases. It consists of three enzymes: pyruvate dehydrogenase, dihydrolipoyl transferase, and dihydrolipoamide dehydrogenase. It also requires five coenzymes: thiamine pyrophosphate, lipoamide, coenzyme A, FAD, and NAD^+.

Tricarboxylic Cycle

What is the function of the TCA cycle?

The two major functions of the TCA cycle are in energy production and biosynthesis.

Where is the TCA cycle located in the cell?

The TCA cycle is located in the mitochondrion (Figure 10.1).

What is the source of the acetyl-CoA used as the starting material for the TCA cycle?

Acetyl-CoA that is used by the TCA cycle may come from pyruvate via the reaction catalyzed by pyruvate dehydrogenase. Acetyl-CoA may also come from the oxidation of fatty acids that occurs in the mitochondrion. Finally, acetyl-CoA may be derived from the degradation of some amino acids.

What is the committed step in the TCA cycle?

The reaction catalyzed by isocitrate dehydrogenase is the committed step of the TCA cycle. It is the first irreversible reaction generating reduced NADH and releasing CO_2. Isocitrate dehydrogenase is also important in the regulation of the TCA cycle. It is inhibited under energy-rich conditions by high levels of NADH and ATP and is activated by NAD^+ and ADP.

What is the consequence of high energy inhibition of isocitrate dehydrogenase?

Inhibition of isocitrate dehydrogenase after a carbohydrate meal results in the accumulation of citrate because the step before isocitrate dehydrogenase in the TCA cycle is reversible and favors formation of citrate from isocitrate.

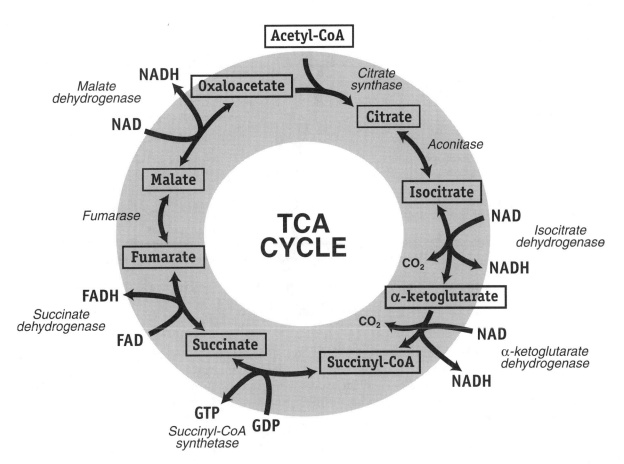

Figure 10.1 The tricarboxylic (TCA) cycle. To initiate the cycle, acetyl-CoA donates its acetyl group to the oxaloacetate to form the six-carbon citrate. This six-carbon compound is committed to the TCA cycle at the isocitrate dehydrogenase step with the loss of CO_2 and production of the five carbon α-ketoglutarate molecule. Although the TCA cycle is central to energy-production, it also plays an important role in providing four and five carbon intermediates, which serve as precursors for other metabolic products.

Accumulation of citrate results in its export to the cytosol, where it activates acetyl-CoA carboxylase and is cleaved by cytosolic citrate lyase to yield oxaloacetate and acetyl-CoA, the latter being the substrate for acetyl-CoA carboxylase and fatty acid biosynthesis.

How does the shuttling of citrate from the mitochondrion to the cytosol also provide for cytosolic NADPH for fatty acid biosynthesis?

The oxaloacetate that results from the cleavage of cytosolic citrate is acted on by malic dehydrogenase to yield malate. The malate is further metabolized by malic enzyme to yield pyruvate that can re-enter the mitochondrion and be converted to oxaloacetate by pyruvate carboxylase. As a result of this cycle, acetyl-CoA equivalents are shuttled out of the mitochondrion in the form of citrate, released in the cytosol, whereas the remainder of the citrate molecule is shuttled back into the mitochondrion to regenerate oxaloacetate. In addition, shuttling results in the conversion of cytosolic reducing equivalents from NADH to NAPH, a form required for fatty acid biosynthesis (Figure 10.2).

What does α-ketoglutarate dehydrogenase have in common with pyruvate dehydrogenase?

α-Ketoglutarate dehydrogenase catalyzes the oxidative decarboxylation of α-ketoglutarate to yield NADH, CO_2, and succinyl-CoA. This enzyme contains three subunits that catalyze similar reactions using the same cofactors as in pyruvate dehydrogenase.

Does α-ketoglutarate dehydrogenase have any other metabolic significance that relates to TCA cycle function?

It is at this point in the TCA cycle, a second carbon is lost in the form of CO_2. Thus, on a net basis, the two carbons contributed by acetyl-CoA at the beginning of the cycle are lost. This, again, explains why acetyl-CoA derived from fatty acid degradation cannot be used to synthesize glucose. It is also important to note that α-ketoglutarate is the keto acid of glutamate and the entry point for the carbon backbone of glutamate and its derivatives for energy production.

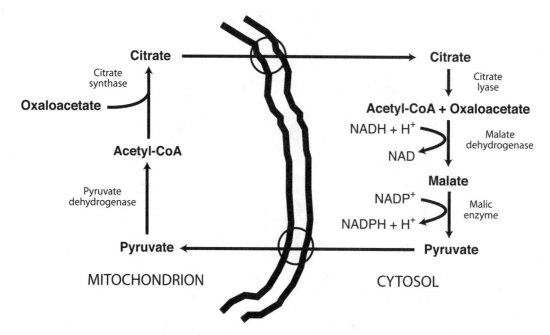

Figure 10.2 The shuttling of citrate from the mitochondrion. Acetate is shuttled out of the mitochondria as citrate. Citrate passes through the inner membrane by a citrate transporter. In the cytosol, citrate is cleaved by citrate lyase, which results in the cytosolic generation of acetyl-CoA. Since oxaloacetate cannot return to the mitochondrial matrix directly, malic enzyme oxidizes the oxaloacetete to pyruvate, generating cytosolic NADPH and pyruvate. Pyruvate can be taken up by the mitochondrion and used by pyruvate dehydrogenase.

What step in the TCA cycle catalyzes substrate-level phosphorylation?

The conversion of succinyl-CoA to succinate catalyzed by succinyl-CoA synthase.

How does the level of oxaloacetate affect TCA cycle function?

Oxaloacetate has a catalytic role in stimulating the entry of acetyl-CoA into the TCA cycle. During fasting, when levels of ATP and NADH, derived from fat metabolism, are increased in the mitochondrion, the increase in NADH shifts the malate:oxaloacetate equilibrium toward malate, which is exported to the cytosol for gluconeogenesis. As a consequence, acetyl-CoA derived from fatty acid oxidation is directed toward ketone body formation by the lack of oxaloacetate, thus promoting ketogenesis during fasting.

Oxidative Phosphorylation

How are reduced coenzymes from the TCA cycle used to produce ATP?

The free energy from the oxidation of reduced coenzymes is used to pump protons to the outside of the inner mitochondrial membrane. These protons re-enter the mitochondrion through an ATP synthase complex. In this way, the proton gradient across the inner mitochondrial membrane drives the synthesis of ATP.

How is the free energy from oxidation of reduced coenzymes derived?

Electrons are transferred from reduced coenzymes through the electron transport chain located in the inner mitochondrial membrane. This produces, in effect, an electrical current used to pump protons to the outside of the inner mitochondrial membrane, resulting in an electrical potential across the membrane.

What is the electron transport chain or respiratory chain?

The electron transport chain (respiratory chain) consists of several large protein complexes and two small, independent components: ubiquinone and cytochrome c. Electrons enter the transport system from NADH (complex I) or $FADH_2$ (complex II) and are conducted to a small lipophilic molecule, ubiquinone (coenzyme Q). At this point, the common electron transport pathway begins, which consists of complex III, cytochrome c, and complex IV. Electrons are conducted through this system to oxygen, and the free energy changes drive the transport of protons from the matrix to the intermediate space of the mitochondrion. Protons are pumped from the matrix at complexes I, III, and IV. The final acceptor of electrons is molecular oxygen, which is reduced to water (Figure 10.3).

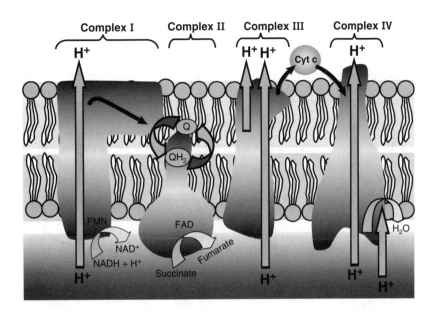

Figure 10.3 The mitochondrial respiratory chain. Electrons pass through a chain of carriers arranged asymmetrically in the inner mitochondrial membrane. Electron flow is accompanied by proton transfer across the inner membrane producing a chemical gradient. Protons re-enter the matrix through a proton-specific channel, which in turn drives the synthesis of ATP.

Why is the yield of ATP from NADH higher than from $FADH_2$?

If electron transport begins with an electron pair from NADH, approximately 3 moles of ATP can be synthesized because protons are pumped from complexes I, III, and IV. In contrast, $FADH_2$ yields about 2 moles of ATP because no protons are pumped from complex II. Only complex III and complex IV pump protons as a result of $FADH_2$ oxidation, thus the lower yield of ATP.

How are reduced equivalents from cytosolic-reduced coenzymes made available to the electron transport system in the mitochondrion?

There are two shuttle systems that transfer the cytosolic reducing equivalents to the mitochondrion. These are the glycerol phosphate shuttle (Figure 10.4) and malate–aspartate shuttle (Figure 10.5).

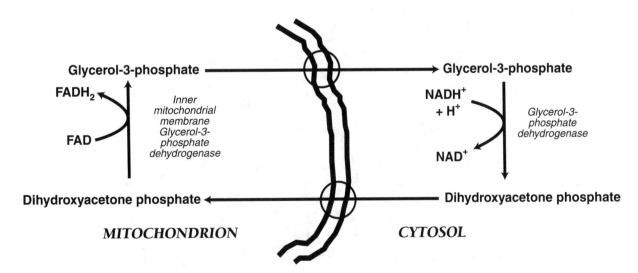

Figure 10.4 Glycerol phosphate shuttle. In skeletal muscle and brain tissue, the glycerol phosphate shuttle provides an alternative means of moving NADH from the cytosol to the mitochondrial matrix.

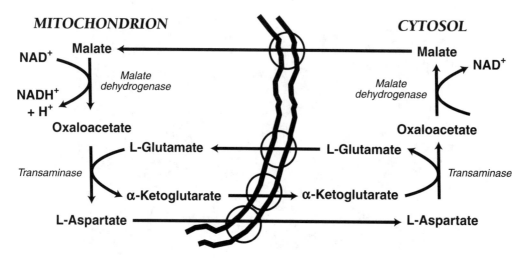

MITOCHONDRION   **CYTOSOL**

Figure 10.5   Malate-apartate shuttle. This shuttle provides the major pathway for transporting reducing equivalents from cytosolic NADH into the mitochondrial matrix of liver, kidney, and heart tissue.

How is the synthesis of ATP coupled to proton transport?

The synthesis of ATP, according to the chemiosmotic hypothesis, is driven by the flux of protons back into the matrix along the electrochemical gradient through ATP synthase complex.

What are the basic features of the ATP synthase complex that allow for its unique function?

ATP synthase is also known as F_0F_1-ATPase and consists of two major complexes. The inner membrane component, termed F_0 because of its sensitivity to oligomycin, contains the proton channel and a stalk piece through which protons can flow back into the mitochondrion. The second, F_1-ATP synthase complex, is bound to F_0 through the stalk, projecting into the matrix. It consists of a central γ-subunit surrounded by alternating α- and β-subunits with a stoichiometry of α_3/β_3. There are three nucleotide-binding sites, located mostly on the β-subunits. The central γ-subunit physically rotates in response to the proton flux. Protons induce rotation of the γ-subunit, which is coupled with the conversion of ADP + P_i to ATP (Figure 10.6).

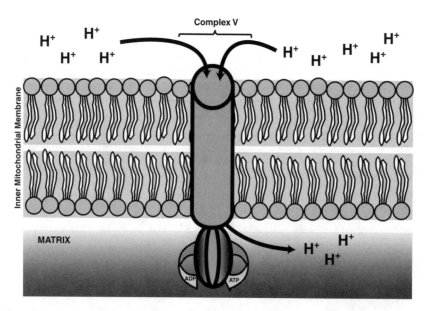

Figure 10.6   The mitochondrial ATP synthase complex. The mitochondrial proton gradient causes the ATP synthase to convert Pi and ADP to ATP.

What is the significance of the P:O ratio?

The P:O ratio is a measure of the number of high-energy phosphate bonds synthesized per atom of oxygen consumed. For example, oxidation of NADH yields about three ATP per oxygen consumed, whereas $FADH_2$ yields about two.

What is an uncoupler of oxidative phosphorylation?

Uncouplers allow transport of protons back into the mitochondria, bypassing the ATP synthase. A common chemical uncoupler is 2,4-dinitrophenol that is hydrophobic and facilitates the movement of protons across the inner mitochondrial membrane.

What would be the result of a natural uncoupler?

In brown adipose tissue, there are natural protein uncouplers whose expression is regulated. The result of this controlled uncoupling of oxidative phosphorylation is to generate heat instead of ATP, and this serves to maintain body temperature.

CHAPTER

11 Fatty Acid Metabolism

Fatty Acids

What is the metabolic significance of fatty acids?

Fatty acids are taken up by cells, where they may serve as fuels for energy production, as precursors in the synthesis of other compounds, and in the liver as substrates for ketone body synthesis.

What are the essential features of a fatty acid?

The essential features of a fatty acid are a long hydrocarbon chain terminating in a carboxylic acid group.

What distinguishes saturated from unsaturated fatty acids?

Unsaturated fatty acids have doubly bonded carbons where the double bond in naturally occurring fatty acids is in the *cis* configuration. The polyunsaturated fatty acids have their doubly bonded carbons spaced by one methyl carbon where this pattern may be repeated to yield fatty acids with multiple double bonds.

Dietary Fat

What are triglycerides?

Triglycerides are composed of fatty acids esterified through each of the three hydroxyl groups of the glycerol backbone (Figure 11.1).

How are triglycerides hydrolyzed in the intestine?

Pancreatic lipase binds to emulsified fat from food with the aid of a colipase. Colipase is a small protein, secreted by the pancreas as an inactive precursor that is activated by trypsin in the duodenum. Pancreatic lipase then hydrolyzes triglycerides to free fatty acids and 2-monoacylglycerol.

Why are triglycerides first hydrolyzed in the intestinal lumen and then resynthesized in the intestinal mucosal cell?

Free fatty acids and monoacylglycerides are slightly water soluble and can diffuse to the intestinal mucosal cell for absorption. Triglycerides cannot be absorbed.

What role do bile salts play in fat digestion?

The absorption of fatty acids and 2-monoacylglycerol requires the presence of bile salts that act as emulsifiers in the small intestine. Bile salts form micelles that acquire fatty acids, 2-monoacylglycerol, and other lipids. From the

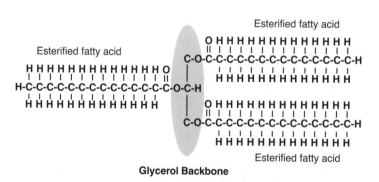

Figure 11.1   Structure of a triglyceride. Triacylglyerol (triglyceride) is made up of three fatty acids esterified to the glycerol backbone.

mixed micelles, fatty acid and 2-monoacylglycerol diffuse to the brush borders and enter the mucosal cells, probably by passive diffusion across the microvillar membrane.

What condition results from failure to digest triglycerides?

The condition is called steatorrhea and can result from pancreatic failure, a lack of bile salts, or extensive intestinal diseases. Steatorrhea is characterized by bulky, fatty stools that contain undigested triglycerides. Fat malabsorption can be effectively treated by enzyme replacement therapy.

How are triglycerides resynthesized in the intestinal mucosal cells?

After free fatty acids and monoacylglycerides are absorbed, they are activated by acyl-CoA synthetase to acyl-CoA derivatives in the intestinal mucosal cells. Triglycerides are resynthesized from 2-monoacylglycerol and the acyl-CoA derivatives.

How are the triglycerides from intestinal mucosal cells delivered to other tissues?

The protein component of chylomicrons is synthesized at the rough endoplasmic reticulum, and the lipids are added in the smooth endoplasmic reticulum and Golgi. Therefore, chylomicrons are a type of lipoprotein that consists of 98% to 99% lipid and 1% to 2% protein and are secreted from the intestinal mucosal cells into the extracellular space, collected by local lymph vessels, and transported to the left brachiocephalic vein and into the blood by the thoracic duct.

How are the triglycerides used by tissue?

Lipoprotein lipase on the surface of capillary endothelium in adipose, skeletal muscle, myocardium, lactating mammary gland, spleen, lung, kidney, and the aorta binds chylomicrons and hydrolyzes the triglycerides to free fatty acids and 2-monoacylglycerol. Most of the fatty acid and monoacylglyceride released by the action of lipoprotein lipase diffuses directly into the tissue. A small percentage also becomes noncovalently bound to serum albumin to be carried in the blood to other sites. The lipoprotein lipase is not synthesized in liver or brain tissue. There are also different isoenzymes of lipoprotein lipase that affect its activity in tissue. For example, fed conditions increase the level of the isoenzyme of lipoprotein lipase in adipose tissue, whereas the myocardial isoenzyme is constitutively expressed, but has a K_m for chylomicrons 10 times lower than the isoenzyme expressed in adipose tissue.

Fat Storage

What role does adipose tissue play in the storage of fat?

After a meal, most of the fatty acid for triglyceride synthesis in adipose tissue is derived from the action of lipoprotein lipase on chylomicrons. Approximately 90% of the fresh weight of adipose tissue is triglyceride, and triglycerides can be resynthesized from absorbed fatty acids that have been activated in the adipose cell by the enzyme thiolase to form the acyl-CoA derivative. In adipose tissue, the backbone used to assemble the triglyceride is derived from glycerol-3-phosphate produced by glycerol phosphate dehydrogenase in the glycolytic pathway. It is important to note that glycerol phosphate does not come from the reaction of glycerol kinase, a reaction important in liver but not adipose tissue.

Under fed conditions, what reactions are regulated that in turn stimulate dietary fat storage in adipose tissue?

Insulin plays a key role in promoting fat storage by stimulating glycolysis, which in turn increases the level of glycerol phosphate available for triglyceride synthesis. Insulin also increases the activity of lipoprotein lipase activity in adipose tissue and thus the supply of fatty acid from chylomicrons. Finally, insulin increases the level of glycerol-phosphate-acyl transferase in adipose tissue. This enzyme catalyzes the addition of the first fatty acid to glycerol phosphate in triglyceride synthesis. The lipolytic hormones are antagonized by insulin.

Under fasting conditions, what reactions are regulated that result in the release of stored fat from adipose tissue?

The process of triglyceride hydrolysis is called lipolysis, which involves the sequential removal of the three fatty acids from glycerol. Hormone-sensitive lipase catalyzes the removal of the first fatty acid from the glycerol backbone and is rate-limiting. Hormone-sensitive lipase is activated by phosphorylation by cAMP-dependent protein kinase A. Catecholamines are the most important of the lipolytic signaling agents that act through β-adrenergic receptors, cAMP, and protein kinase A. Norepinephrine released from sympathetic nerve terminals in adipose tissue appears to be the significant signaling agent in responding to food deprivation or cold exposure (Figure 11.2).

Do other hormones facilitate lipolysis?

Leptin, glucocorticoids, growth hormone, and thyroid hormone stimulate lipolysis, but not through cAMP-dependent protein kinase A, but rather by increasing the synthesis of lipolytic proteins.

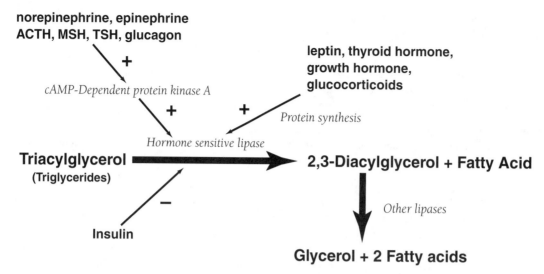

Figure 11.2 Hormonal regulation of lipolysis in adipose tissue. Hormone-sensitive lipase hydrolyzes triacylglycerol in adipocytes, releasing fatty acids. The fatty acids pass from the adipocyte into the blood, where they bind to serum albumin. Hormones such as epinephrine and glucagon, secreted in response to low blood glucose levels, activate the enzyme adenylyl cyclase in the adipocyte membrane, which produces the intracellular second-messenger cAMP. cAMP activates cAMP-dependent protein kinase, which phosphorylates perilipin, which causes hormone-sensitive lipase in the cytosol to move to the lipid droplet surfaces in the adipocyte to hydrolyze triacylglycerides.

What do the phenotypic features of Cushing's syndrome suggest about the lipolytic response to catecholamines?

In patients with Cushing's syndrome, where there is an excess of glucocorticoids produced as a result of a pituitary tumor, the expected reduction in adipose tissue is in the extremities. Adipose tissue in the abdomen and neck do not respond, suggesting regional adipose variation in response to glucocorticoid signaling.

What form of lipid does adipose tissue release?

Unlike liver and intestine, adipose tissue does not release particulate lipid in the form of lipoproteins, but rather releases "free" (unesterified) fatty acids, which bind noncovalently and reversibly to serum albumin for transport. Albumin bound fatty acids have a rapid turnover, with a plasma half-life of only 3 minutes.

Fatty Acid Biosynthesis

In humans, where does endogenous fatty acid synthesis occur?

Endogenous fatty acid synthesis is important in only a few tissues, such as liver and lactating mammary gland.

Where does fatty acid synthesis take place in the cell?

Fatty acid synthesis takes place in the cytoplasm.

What is the major source of acetyl-CoA for fatty acid biosynthesis?

Glucose is the major source of acetyl-CoA for fatty acid biosynthesis. Glucose is first degraded to pyruvate by aerobic glycolysis (in the cytoplasm). Pyruvate dehydrogenase (in the mitochondria) oxidatively decarboxylates pyruvate forming acetyl-CoA and other products. Acetyl-CoA can then serve as a substrate for citrate synthesis, and citrate, in turn, can be transported out of the mitochondria and split in the cytoplasm to generate cytoplasmic acetyl-CoA for fatty acid synthesis.

What is the significance of the thioester bond linking the acetyl group to the CoA group in the acetyl-CoA structure?

The thioester bond linking the acetyl group to CoA is a "high-energy" bond.

What is the regulated step in fatty acid biosynthesis?

The rate-limiting and regulated step in fatty acid biosynthesis is catalyzed by the enzyme acetyl-CoA carboxylase, where acetyl-CoA is carboxylated in an energy-requiring reaction to produce malonyl-CoA.

$$\text{Acetyl-CoA Carboxylase}$$
$$(Biotin)$$
$$\text{Acetyl-CoA} + CO_2 \rightarrow \text{Malonyl-CoA}$$

What are the significant requirements of acetyl-CoA carboxylase?

Acetyl-CoA carboxylase is a biotin-requiring enzyme, where biotin acts as a prosthetic group. It is bound to acetyl-CoA carboxylase through an amide link of an ϵ-amino group of a lysine residue.

How is the synthesis of fatty acids regulated relative to fed and fasting conditions?

In part, fatty acid synthesis is regulated by changes in the level of acetyl-CoA carboxylase, fatty acid synthase, ATP-citrate lyase, and glucose-6-phosphate dehydrogenase in the liver. Levels are increased during carbohydrate feeding and decreased during fasting. For example, a fat-free, carbohydrate-based diet is the most potent inducer of these enzymes. The short-term regulation of fatty acid synthesis occurs at the rate-limiting step, acetyl-CoA carboxylase. This enzyme exists in an inactive monomeric form and an active polymeric form. Citrate is the most important allosteric regulator and stabilizes the active polymeric form, whereas palmitoyl-CoA dissociates the complex into the inactive monomers.

What effect do hormones have on the regulation of fatty acid biosynthesis?

Insulin stimulates acetyl-CoA carboxylase, whereas glucagon and epinephrine inhibit it. cAMP protein kinase A mediates the glucagon and epinephrine response by phosphorylation of acetyl-CoA carboxylase, thereby inhibiting its activity (Figure 11.3).

Why are there two isoforms of acetyl-CoA carboxylase?

Malonyl-CoA, the product of the acetyl-CoA carboxylase (ACC) catalyzed reaction, not only serves as the substrate for fatty acid biosynthesis, but also as a signal molecule for metabolic control of fatty acid β-oxidation in skeletal muscle and insulin secretion in pancreatic β-cells. Two isoforms of ACC have been identified (ACC_α and ACC_β). These two isoforms are encoded by separate genes and display distinct tissue distribution. ACC_α is found mainly in liver and adipose tissue, where lipogenesis is active. In contrast, ACC_β is predominant in skeletal muscle and heart, where the β-oxidation of fatty acids serves as the main energy source. Therefore, the level of malonyl-CoA generated by ACC_β functions as an important signal in regulating mitochondrial fatty acid β-oxidation through the inhibition

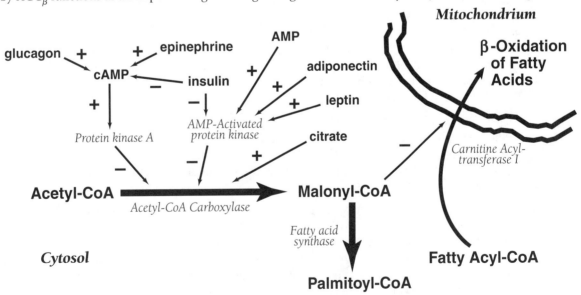

Figure 11.3 Regulation of fatty acid metabolism. Fatty acids in the cytosol have two possible routes. One is β-oxidation by enzymes in the mitochondria. The second is conversion into triacylglycerols and phospholipids by enzymes in the cytosol. The pathway taken depends on the rate of transfer of the long-chain fatty acid into mitochondria. The three-step process (carnitine shuttle) by which fatty acyl groups transfer from the cytosol to the mitochondrial matrix is regulated by the first intermediate of fatty acid synthesis, malonyl-CoA.

of carnitine palmitoyl-CoA transferase I (CPT1). The activity of ACC_β in skeletal muscle is essentially regulated by phosphorylation and dephosphorylation and not by changes in the level of enzyme. For example, sympathetic nerve stimulation or exercise inhibits ACC_β activity by phosphorylation, which in turn results in the increase of fatty acid β-oxidation in skeletal muscle.

What is the first step catalyzed by fatty acyl synthase?

An acetyl group is transferred from acetyl-CoA to the -SH group of the condensing enzyme domain of fatty acyl synthase, forming acetyl-CE. The reaction is catalyzed by the acyltransferase activity of fatty acyl synthase.

How is the malonyl group transferred to the acyl carrier peptide on fatty acyl synthase?

During fatty acid synthesis, the incoming two-carbon fragment is introduced as the three-carbon malonyl group. It is added to the -SH group of the acyl-carrier peptide domain of fatty acid synthase. In a subsequent reaction, a carbon is lost as a bicarbonate ion.

What product is released during the condensation of an acyl group with a malonyl group that can be used by acetyl-CoA carboxylase?

$$\underset{\substack{\text{Malonyl acyl carrier} \\ \text{peptide}}}{HO_2C\text{-}CH_2\text{-}\overset{\displaystyle O}{\overset{\displaystyle \|}{C}}\text{-}S\text{-}ACP} + \underset{\substack{\text{Acetyl condensing}}}{CH_3\text{-}\overset{\displaystyle O}{\overset{\displaystyle \|}{C}}\text{-}S\text{-}cys\text{-}CE} \rightarrow \underset{\substack{\beta\text{-Keto acyl carrier} \\ \text{peptide}}}{CH_3\text{-}\overset{\displaystyle O}{\overset{\displaystyle \|}{C}}\text{-}CH_2\text{-}\overset{\displaystyle O}{\overset{\displaystyle \|}{C}}\text{-}ACP} + HCO_3 + HS\text{-}cys\text{-}CE$$

The acetyl group displaces the carboxyl of the malonyl group, forming a β-ketoacyl group. This reaction is catalyzed by β-ketoacyl acyl carrier peptide synthase. Significantly, the carboxyl group released in the form of bicarbonate regenerates the bicarbonate used earlier in the acetyl-CoA carboxylase reaction.

Where in fatty acid biosynthesis are the reducing equivalents supplied by NADPH used?

$$\underset{\substack{\beta\text{-Keto acyl carrier} \\ \text{peptide}}}{CH_3\text{-}\overset{\displaystyle O}{\overset{\displaystyle \|}{C}}\text{-}CH_2\text{-}C\text{-}ACP} + NADPH + H^+ \rightarrow \underset{\substack{\beta\text{-Hydroxy acyl carrier} \\ \text{peptide}}}{CH_3\text{-}\overset{\displaystyle OH}{\overset{\displaystyle |}{CH}}\text{-}CH_2\text{-}\overset{\displaystyle O}{\overset{\displaystyle \|}{C}}\text{-}S\text{-}ACP} + NADP^+$$

Reduction of β-ketoacyl acyl carrier peptide of the fatty acid synthase complex.

Where do the reducing equivalents, NADPH, come from for fatty acid biosynthesis?

The reducing equivalents, NADPH, come from the oxidative side of the hexose monophosphate shunt and from malic enzyme that converts malate to pyruvate. Theoretically, the reaction catalyzed by malic enzyme could supply half of the NADPH required for fatty acid synthesis.

How are fatty acids longer than palmitate produced?

Fatty acid synthase complex produces only palmitate, a saturated, 16-carbon fatty acid. Chain elongation is possible through an elongations system found in both the endoplasmic reticulum and the mitochondria. The mitochondrial system prefers fatty acids with fewer than 16 carbons, whereas the endoplasmic reticulum system works best with palmitate. Chain elongation occurs with unsaturated as well as saturated fatty acids.

Because only 50% of the body's fatty acids are saturated, how are the monounsaturated and polyunsaturated fatty acids synthesized?

The desaturation of fatty acids is catalyzed by a desaturase system found in the endoplasmic reticulum of liver and other organs. The first double bond is introduced at carbon 9 of palmitic or stearic acid, producing palmitoleic or oleic acid, $\omega7$ and $\omega9$ classes of polyunsaturated fatty acids, respectively. Oleic acid is the most abundant of the body's unsaturated fatty acids. Additional double bonds can be introduced between the first double bond and the carboxyl group, but not beyond carbon 9. Therefore, many polyunsaturated fatty acids can be synthesized from palmitic or stearic acid by a combination of desaturation and chain elongation.

Why are some polyunsaturated fatty acids essential?

Linoleic acid ($18:2^{\Delta 9,12}$) and linolenic acid ($18:3^{\Delta 9,12,15}$), the parent compounds of the Ω 6 and Ω 3 classes of polyunsaturated fatty acids, respectively, cannot be synthesized in the human body and therefore are essential. Essential fatty acid deficiency characterized by dermatitis and poor wound healing has been observed in patients kept on total parenteral nutrition for long time periods and patients suffering from severe fat malabsorption

Fatty Acid Oxidation

In what compartment of the cell are fatty acids degraded?

The mitochondrion is the principal site of fatty acid oxidation.

How do fatty acids get into the mitochondria?

Transport of the long-chain fatty acids across the inner mitochondrial membrane requires the enzyme carnitine-palmitoyltransferase (carnitine acyl-transferase). There are two transferases, designated I and II, that are present on the outer and inner surface of the inner mitochondrial membrane, respectively. Short- and medium-chain fatty acids diffuse passively across the membrane and are subsequently activated in the mitochondria.

How is energy derived from the oxidation of fatty acids?

The principal fate of fatty acyl-CoA in the mitochondrion is β-oxidation, where each cycle of β-oxidation shortens the fatty acyl-CoA substrate by two carbons to produce acetyl-CoA. Acetyl-CoA, NADH, and FADH that result from the oxidation of fatty acids in the mitochondria can be used by the TCA cycle and respiratory chain, respectively.

How is the oxidation of fatty acids regulated?

Basically, the oxidation of fatty acids is regulated based on supply. The supply of fatty acids is itself regulated at two levels. First, the availability of fatty acids by tissues is proportional to the plasma level of fatty acids, and therefore, this level is regulated by hormone-sensitive lipase in adipose tissue. Second, intracellular β-oxidation is regulated by the uptake of fatty acyl-CoA into the mitochondrion. Because carnitine acyl-transferase is allosterically inhibited by malonyl-CoA, uptake of fatty acids into the mitochondrion is regulated by fatty acid synthesis. Increased malonyl-CoA levels in the cell as a result of increased fatty acid synthesis would inhibit fatty-acyl CoA uptake for mitochondrial β-oxidation (Figure 11.3).

Is the oxidation of fatty acids important in all tissues?

Some tissues, including nervous tissue and red blood cells, do not oxidize fatty acids. Other tissues, including muscle and liver, derive most of their energy from fatty acids, particularly during the fasting. The importance of this difference is that a deficiency in the ability to oxidize fatty acids would be most pronounced in the function of liver and muscle.

Is there an example in which a deficiency has been observed in fatty oxidation?

Deficiencies in the enzymes within the β-oxidation pathway are known but are quite rare. Liver and muscle in particular can suffer from a deficiency in carnitine that results from defects of carnitine biosynthesis, defective transport into cells, or excessive renal excretion. Patients with carnitine deficiency present with muscle weakness and muscle cramps on exertion. Tissue biopsy of muscle and liver shows an unusual abundance of fat droplets, to the extent that some patients develop fatty degeneration of the liver. These fatty droplets develop because the excess acyl-CoA that results from carnitine deficiency cannot be imported in the mitochondrion and is therefore diverted into triglyceride synthesis.

Under what circumstances might there be an excessive excretion of carnitine by the kidney?

Carnitine is depleted in many organic acidurias, including methylmalonic aciduria that results from a B_{12} deficiency, because the accumulating acids form carnitine esters, which are excreted in the urine. Carnitine esters can be formed from CoA-thioesters of many organic acids in reactions analogous to the carnitine–palmitoyltransferase reaction and represent, by themselves, a mechanism to detoxify potentially hazardous organic acids.

Why does hypoketotic hypoglycemia develop in patients with carnitine deficiency during extended periods of fasting?

Gluconeogenesis requires energy in the form of ATP and GTP, and fatty oxidation is the only important energy source for the fasting liver. Therefore, any defect in the main sequence of β-oxidation prevents gluconeogenesis. Ketogenesis is impaired as well, because the supply of acetyl-CoA from β-oxidation is blocked. Importantly, the lack of ketone bodies adds to the effect of the hypoglycemia because the brain depends entirely on a combination of glucose and ketone bodies during periods of fasting.

Do unsaturated fatty acids require an alternative pathway in order to be degraded?

Unsaturated fatty acids require modification of their double bonds before β-oxidation. For example, the *cis* double bonds of unsaturated fatty acids must be isomerized during oxidation to the *trans* form, an intermediate of β-oxidation.

What purpose does α-oxidation of fatty acids serve?

This is considered a minor oxidative pathway that shortens the fatty acid by one carbon at a time. The significance of this pathway is illustrated by Refsum's disease, a rare recessively inherited disorder with a defect in α-oxidation. These patients accumulate excessive quantities of phytanic acid, a constituent of chlorophyll in green vegetables, in their body lipids that cannot be oxidized by β-oxidation without prior modification because the β-carbon is methylated. In normal cases, phytanic acid degradation is initiated by α-oxidation, followed by β-oxidation.

Why does a deficiency in peroxisomes (Zellweger's syndrome) cause the accumulation of very long-chained fatty acids?

Very long-chain fatty acids (> 20 carbons) are poor substrates for mitochondrial β-oxidation. Peroxisomal β-oxidation uses essentially the same reaction sequence as the mitochondrial system except for the very first reaction, which is catalyzed by H_2O_2-producing flavoproteins and can proceed only to the stage of octanoyl-CoA. Thus, peroxisomes provide a system that shortens very long-chained fatty acids to a point where they can be further oxidized by the mitochondrial β-oxidation system.

Ketone Bodies

What are ketone bodies?

The term *ketone bodies* refers to three biosynthetically related products: acetoacetate, β-hydroxybutyrate, and acetone.

Where are ketone bodies formed?

Acetoacetate and β-hydroxybutyrate are formed from acetyl-CoA in liver mitochondria via a pathway that produces HMG-CoA as an intermediate. The first part of this pathway is identical to that which forms HMG-CoA as an intermediate in cholesterol biosynthesis. The critical difference, however, is that ketone body synthesis occurs in mitochondrion, whereas cholesterol biosynthesis occurs in the cytosol, involving enzymes associated with the smooth ER (Figure 11.4).

How are ketone bodies used by the body?

The liver releases ketone bodies into the blood for transport to other tissues where they serve as an important energy source. Importantly, the liver itself cannot use ketone bodies. Acetone, which is formed by nonenzymatic decarboxylation of acetoacetate, apparently serves no biological function, and most is exhaled through the lungs. In diabetic ketoacidosis, there is a sufficient level of acetone in the patient's breath to be recognized and used diagnostically.

How is the synthesis of ketone bodies regulated?

Fasting, fat feeding, and insulin insufficiency increase the synthesis of liver mitochondrial HMG-CoA synthase, the rate-limiting enzyme in ketogenesis. During fasting, there is a large increase in β-oxidation. The acetyl-CoA formed by β-oxidation is not used for oxidation by the TCA cycle in the liver because the liver receives most of its energy from the respiration of the NADH and $FADH_2$ formed during β-oxidation. In addition, oxaloacetate levels are most often limiting for conjugation with acetyl-CoA, and thus, the excess acetyl-CoA from fatty acid oxidation is shunted into ketone body synthesis. The ketone bodies are transported out of the liver to supply energy needs of extrahepatic tissues. Remember, the liver cannot use ketone bodies.

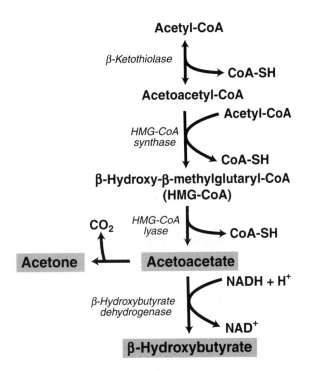

Figure 11.4 Biosynthesis of ketone bodies in the liver. The three ketone bodies are designated by bold letters in a gray background. Acetone is believed to be formed by the non-enzymatic decarboxylation of acetoacetate, which occurs at very low levels.

12 Membrane Lipid Metabolism

Metabolism of Phosphoglycerides and Sphingolipids

What are the general classes of lipids that make up cellular membranes?

There are three chemical classes of membrane lipid: the phosphoglycerides, the sphingolipids, and the sterols.

Where are these lipids synthesized?

Most cells are able to synthesize these lipids, but there is also a significant level of tissue exchange of membrane lipids that takes place via plasma lipoproteins. Both phosphoglycerides and sphingolipids can be synthesized *de novo* in all cells, except the red blood cell.

Where does the glycerol phosphate backbone in the phosphatidic acid structure come from?

Glycerol-3-phosphate comes from dihydroxyacetone phosphate in the glycolytic pathway. The addition of two acyl groups represents the same reaction sequence used for the synthesis of triglycerides in tissue other than the intestine.

How is phosphatidic acid converted to phosphoglycerides?

There are two pathways for the *de novo* synthesis of the phosphoglycerides. In one pathway, phosphatidic acid is activated in the form of CDP-diacylglycerol; in the other, the alcohol is activated. In both cases, CTP is used to activate one of the substrates providing the energy for the coupling reaction. As a result of coupling, the phosphoanhydride bond ("high-energy bond") of CDP derivative is cleaved, and one of the phosphate groups of CDP is incorporated in the product. The first pathway is used for the synthesis of phosphatidylinositol and cardiolipin, and the second pathway is the major source of phosphatidylethanolamine and phosphatidylcholine (Figure 12.1).

How can phosphoglycerides be remodeled after their initial synthesis?

Phosphoglycerides are acted on by phospholipase. These enzymes may act to degrade phosphoglycerides or to allow the exchange of fatty acids in position 1 and 2 by hydrolysis followed by reacylation of a different fatty acid. Acyl transferases that catalyze the addition of the fatty acid direct the specificity of positions 1 and 2 on the glycerol backbone. Position 1 generally has a fatty acid that is saturated, whereas position 2 is generally occupied by an unsaturated fatty acid. Also, the alcohol may be exchanged as well. For example, phosphatidylserine is synthesized mostly by exchange with ethanolamine in human tissues. Moreover, phosphatidylethanolamine can be made by decarboxylation of phosphatidylserine, and phosphatidylcholine can be made by methylation of phosphatidylethanolamine.

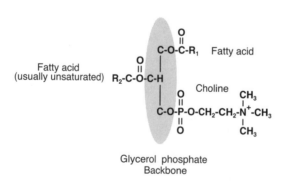

Figure 12.1 Structure of phosphatidylcholine. Phosphatidylcholine is made up of two fatty acids and phosphocholine esterified to the glycerol backbone. The fatty acid esterified to the second position of glycerol is usually unsaturated. Phosphatidylcholine is one of the most abundant phospholipids in most eukaryotic cells.

How do sphingolipids differ from phosphoglycerides?

In sphingolipids, ceramide, consisting of sphingosine and a long-chain fatty acid, is the core structure or backbone rather than glycerol. In addition, the primary hydroxy group at carbon-1 of sphingosine is substituted by phosphocholine in the case of sphingomyelin and by a monosaccharide or oligosaccharide in the glycosphingolipids (Figure 12.2).

Where does the sphingosine come from?

Sphingosine, part of the ceramide backbone of sphingolipids, is synthesized in the endoplasmic reticulum of most cells from palmitoyl-CoA and serine. The fatty acid, in the form of a CoA-thioester, is then attached to form ceramide.

How are the monosaccharide residues introduced from glycosphingolipids?

During the synthesis of the glycosphingolipids, the monosaccharide residues are introduced from their activated precursors. The more complex oligosaccharide chains of the complex glycosphingolipids (globosides and gangliosides) are built by a stepwise addition of monosaccharide units catalyzed by specific transferases.

How do the ABO blood groups relate to glycosphingolipids?

The antigenic variation is caused by the difference in terminal sugar moiety of the oligosaccharide of glycosphingolipids of the erythrocyte membrane. The GalNac (N-acetyl-galactose transferase and galactose transferase) is encoded by two allelic variations of the same gene, known as the A allele and B allele. There is also a third allele of this gene that encodes an inactive form of the enzyme. Therefore, a person may have type O blood if both alleles express the inactive enzyme, or type A blood if one is the A allele, the other the inactive allele, or if both are A alleles. Type B blood results from one B allele, the other inactive, or when both alleles are B. Finally, if an individual has one A allele and one B allele, then they will express the blood type AB.

Why are deficiencies of sphingolipid degrading enzymes grouped under lipid storage diseases?

Lysosomal enzymes catalyze the orderly degradation of sphingolipids. As a result of a deficiency in one of the degradative enzymes, there is a progressive accumulation of a nonsoluble, nondegradable, nonexcretable lipid. The nervous system is seriously affected in all cases because of its high sphingolipid content and, therefore, the necessary turnover. Hepatosplenomegaly is also associated with these diseases because the phagocytic cells in the spleen and liver remove erythrocytes from the circulation, and nondegradable lipid from the red blood cell membrane accumulates in these tissues.

What disease examples fall under the category of lipid storage diseases?

Gaucher's disease (a deficiency in β-glucosidase, also known as glucocerebrosidase) and Tay-Sachs disease (a deficiency in hexosaminidase A) are examples of the more common diseases that result from a deficiency of a specific

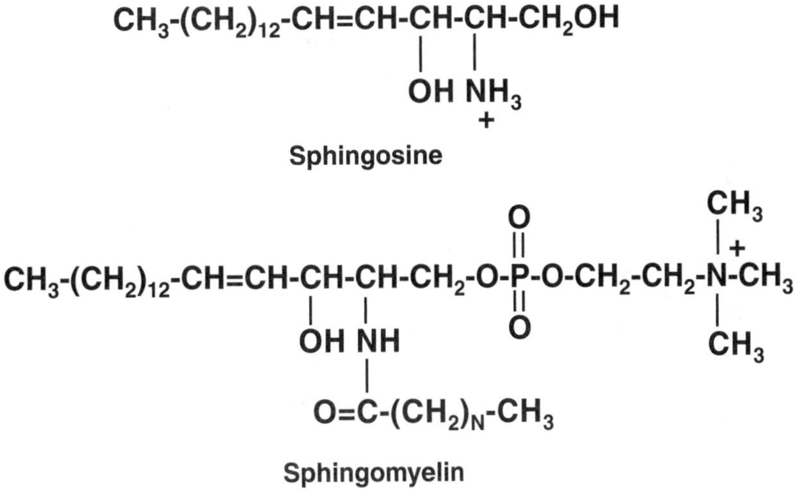

Figure 12.2 Structures of sphingosine and sphingomyelin. Sphingolipids are derived from sphingosine. Sphingomyelin is a common sphingolipid found in the membranes of animal cells.

lysosomal enzyme. Tay-Sachs disease is very rare in most populations, but in Ashkenazi Jews, it occurs with a frequency of approximately 1 in 3,000 births. Although the affected children appear normal at birth, they develop signs of mental and neurologic deterioration within the first year of life. This is accompanied by hepatomegaly, and usually the child dies before the age of 3 years. Tay-Sachs disease results from a complete deficiency of hexosaminidase A (Figure 12.3).

Eicosanoid Metabolism

Eicosanoids can be synthesized from which essential fatty acids?

The eicosanoids are synthesized from C_{20} eicosanoic acids. The C_{20} eicosanoic acids are derived from polyunsaturated essential fatty acid linoleate and α-linolenate or directly from arachidonate and eicosapentaenoate in the diet (Figure 12.4).

What groups of compounds make up the eicosanoids?

The groups are leukotrienes, lipoxins, prostaglandins, and thromboxanes.

What step in the synthesis of eicosanoids is the rate-limiting step?

The availability of substrate, in the form of arachidonic acid, for eicosanoid biosynthesis is highly regulated. This step is catalyzed by phospholipase A_2, which releases arachidonic acid from plasma membrane phospholipids. This step is activated by compounds such as angiotensin II, bradykinin, epinephrine, and thrombin. Corticosteroids act as anti-inflammatory agents by inhibiting this step.

What is the committed step for prostanoid synthesis?

The pathway for the synthesis of prostaglandins and thromboxanes is called the cyclooxygenase pathway, where the committed step is catalyzed by prostaglandin endoperoxidase synthase. This enzyme possesses two separated enzyme activities, named cyclooxygenase and peroxidase.

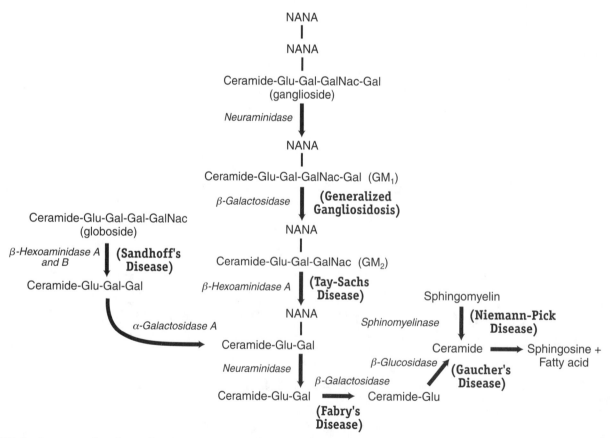

Figure 12.3 Lysosomal pathway for degradation of sphingolipids. Genetic defects in glycosphingolipid catabolism cause breakdown intermediates to accumulate in the nervous system tissue with severe consequences.

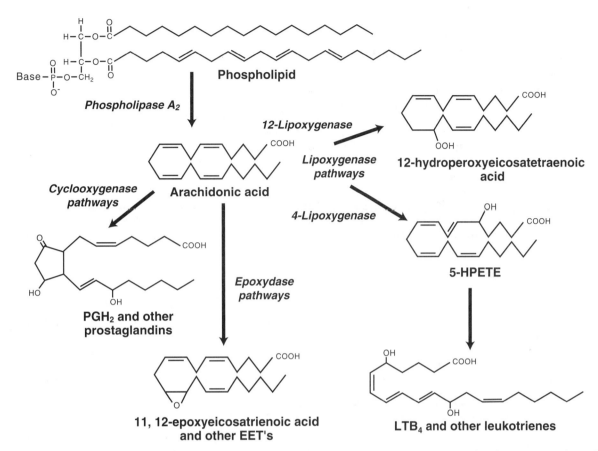

Figure 12.4 Eicosanoids biosynthesis. Eicosanoids are a class of lipids that are distinguished by their potent physiological properties, low levels in tissues, rapid metabolic turnover, and common metabolic origin.

There are two distinct isoforms of cyclooxygenase, termed *COX-1* and *COX-2*. How do these isoforms differ?

COX-1 is constitutively expressed, whereas COX-2 is made only in response to inflammatory mediators such as cytokines.

What are the products of the cyclooxygenase pathway?

The product of the cyclooxygenase pathway is an endoperoxide (PGH) that is subsequently converted to prostaglandins D, E, and F as well as to the thromboxane and prostacyclin. Importantly, a particular cell type produces only one type of prostanoid.

How do aspirin and aspirin-like derivatives affect prostanoid synthesis?

Aspirin inhibits the cyclooxygenase. The aspirin-like derivatives indomethacin and ibuprofen likewise inhibit cyclooxygenase activity. These cyclooxygenase inhibitors have been termed nonsteroidal anti-inflammatory drugs and act to reduce inflammation and provide pain relief by blocking the first step in production of prostaglandins.

What advantage would a selective drug that inhibits only COX-2 have over the currently marketed drugs that preferentially block COX-1?

The major side effects of blocking COX-1 are bleeding and the inflammation of gastric mucosa. Thus, the use of inhibitors selective for COX-2 may make it possible to target inflammatory activity more specifically and avoid the side effects associated with the use of COX-1 inhibitors.

Does a deficiency in essential fatty acids affect prostaglandin synthesis?

Although there is a marked correlation between essential fatty acids and their ability to be converted to prostaglandins, the need for essential fatty acids in membrane formation is also very important and is unrelated to prostaglandin synthesis.

How do prostaglandins affect vascular function?

Platelets release thromboxanes, which cause vasoconstriction and platelet aggregation. In contrast, prostacyclins are produced by blood vessel walls and are potent inhibitors of platelet aggregation. Thus, thromboxanes and prostacyclins are antagonistic.

What is the committed step in leukotriene and lipoxin biosynthesis?

The committed step is catalyzed by enzymes called lipoxygenases. Three different lipoxygenases catalyze the insertion of oxygen into the 5, 12, 15 positions of arachidonic acid. Only 5-lipoxygenase forms leukotrienes. The lipoxins are formed by a combination of lipoxygenases, which introduce more oxygen into the molecule.

What type of tissue produces leukotrienes and lipoxins?

Leukotrienes are formed from eicosanoic acids in leukocytes, mastocytoma cells, platelets, and macrophage in response to immunologic and nonimmunologic stimuli. Similarly, leukocytes also produce lipoxins.

What relationship do leukotrienes have with the anaphylactic response?

The slow-reacting substance of anaphylaxis is a mixture of leukotrienes C_4, D_4, and E_4. The response to these leukotrienes is 100 to 1000 times more potent than histamine or prostaglandins in causing constriction of the bronchial airway musculature.

What role do leukotrienes and lipoxins play in vascular function?

Leukotrienes affect vascular permeability and the attraction and activation of leukocytes. Less is known about the action of lipoxins other than their vascular activity may counter regulatory compounds of the immune response.

Cholesterol Metabolism

What cellular structure in animal cells contains cholesterol?

The body contains approximately 140 g of cholesterol, the majority of which is unesterified cholesterol found in the cellular membrane (Figure 12.5).

What is the source of tissue cholesterol?

A little more than 50% of the cholesterol of the body is synthesized, and the remainder is provided by the diet.

How is dietary source of cholesterol taken up in the body?

Cholesterol esters in the diet are hydrolyzed to cholesterol, which mixes with dietary unesterified cholesterol and biliary cholesterol before absorption from the intestine in the company of other lipids. It then mixes with cholesterol synthesized in the intestines, and together with other dietary lipids, the cholesterol esters are packaged into chylomicrons. Of the cholesterol absorbed, 80% to 90% is esterified to long-chain fatty acids in the intestinal mucosa by the microsomal enzyme acyl-CoA-cholesterol acyl transferase.

What is the composition of a chylomicron?

Chylomicrons are approximately 98% to 99% lipid and 1% to 2% protein. Most of their lipid is triglyceride, but they also contain other dietary lipids such as cholesterol, cholesterol esters, phospholipids, and fat-soluble vitamins. The protein component of the chylomicrons is synthesized at the rough endoplasmic reticulum, and the lipids are added in the smooth endoplasmic reticulum and Golgi of the intestinal mucosa cell.

What happens to the lipid packaged into chylomicrons?

Chylomicrons are secreted into the extracellular space, collected by local lymph vessels, and transported to the left brachiocephalic vein by the thoracic duct. Most of the triglyceride packaged into chylomicrons is removed by extrahepatic lipoprotein lipase attached to heparin sulfate proteoglycans on the surface of the capillary endothelium of

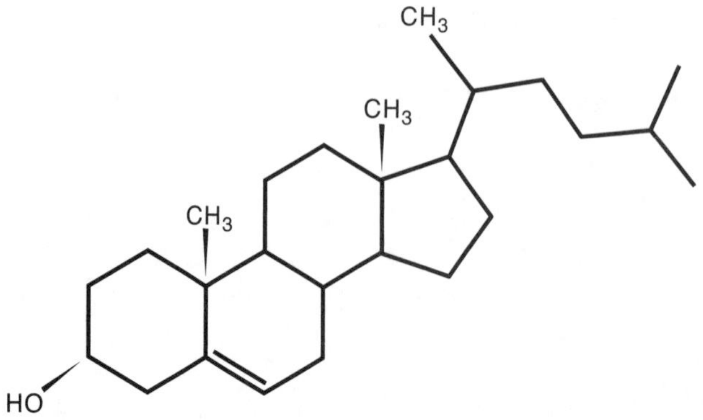

Figure 12.5 Structure of cholesterol. Cholesterol derives all of its carbon atoms from acetate.

adipose tissue, skeletal muscle, the myocardium, lactating mammary gland tissue, spleen, lung, kidney, and aorta but not in liver and brain. The cholesterol-rich remnant particle is then taken up by the liver, and in this way, most of the dietary cholesterol is directed to the liver.

Do all tissues synthesize cholesterol?

Virtually all tissues containing nucleated cells are capable of synthesizing cholesterol; however, some tissue, such as the liver, account for more than 50% of the total synthesis. Some endocrine tissues, such as the adrenal cortex and the corpus luteum, which produce steroid hormones, have high rates of cholesterol synthesis.

How is cholesterol synthesized?

Cholesterol is synthesized from acetyl-CoA. The committed step of the cholesterol synthetic pathway is catalyzed by hydroxy-3-methylglutaryl-CoA reductase (HMG-CoA reductase). Acetyl-CoA condenses to form acetoacetyl-CoA catalyzed by cytosolic thiolase. Then acetoacetyl-CoA condenses with a further molecule of acetyl-CoA catalyzed by HMG-CoA synthase to form HMG-CoA. It is the conversion of HMG-CoA to mevalonate catalyzed by HMG-CoA reductase that is considered the rate-limiting step in the pathway of cholesterol biosynthesis. The pathway for HMG-CoA synthesis involved in cholesterol biosynthesis follows the same sequence of reactions described for the synthesis of ketone bodies. However, ketone body synthesis occurs in the mitochondria, whereas cholesterol biosynthesis is extramitochondrial; thus, the two pathways are distinct. The six-carbon mevalonate is then processed to the branched-chain five-carbon compounds isopentenyl pyrophosphate and dimethylallyl pyrophosphate. Six of the five-carbon units are then assembled to form a 30-carbon compound called squalene, and squalene is then cyclized to lanosterol. Lanosterol is processed further to yield cholesterol (Figure 12.6).

Are there other biologically significant products derived from the cholesterol biosynthetic pathways?

Beside cholesterol and other steroids, isoprenoids are derived from the branched-chain five-carbon compounds isopentenyl pyrophosphate and dimethylallyl pyrophosphate. Included in these isoprenoids is the side chain (10 isoprene units) of ubiquinone (coenzyme Q), a member of the respiratory chain dolichol, which participates in the syn-

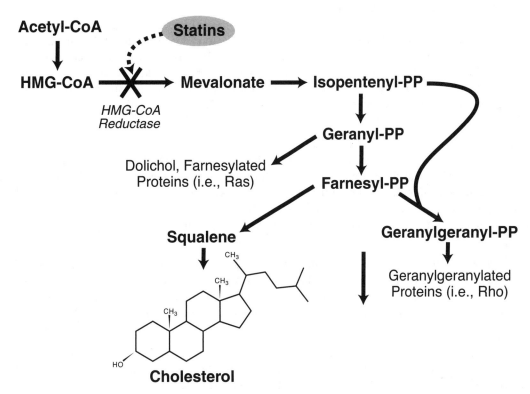

Figure 12.6 Biosynthesis of cholesterol. The first part of the pathway is identical to reactions used in ketogenesis except that it occurs in a different cellular compartment. Cholesterol synthesis occurs in the cytosol and the endoplasmic reticulum, whereas ketogenesis occurs in the mitochondria. As noted, HMG-CoA reductase, which catalyzes the committed reaction in cholesterol biosynthesis, represents a major target for regulation of the overall pathway, as well as a target for pharmacological control.

thesis of N-linked oligosaccharides in glycoproteins, and farnesyl and geranylgeranyl groups, which are used to anchor proteins to membranes.

How is HMG-CoA reductase regulated to affect cholesterol biosynthesis?

Cholesterol biosynthesis is regulated by feedback inhibition; thus, free cholesterol feeds back and inhibits the level of HMG-CoA reductase activity. This effect appears to be mediated not directly by cholesterol, but by a cholesterol metabolite at both the transcriptional and posttranscriptional level. Insulin or thyroid hormones increase HMG-CoA activity, whereas glucagon or glucocorticoids decrease HMG-CoA reductase activity. Cholesterol synthesis is also inhibited by low-density lipoprotein (LDL) cholesterol taken up via LDL receptors.

Is HMG-CoA reductase a site for therapeutic intervention to control serum cholesterol levels by blocking *de novo* biosynthesis?

HMG-CoA reductase inhibitors (statins) such as mevastatin and lovastatin are used to lower serum cholesterol levels. Their mode of action is somewhat indirect though. As a consequence of their inhibitor action on HMG-CoA reductase, they upregulate the LDL receptors in the liver, thus lowering serum LDL cholesterol levels as a consequence of increased uptake by the liver.

Are there other strategies used to lower serum cholesterol levels?

High serum cholesterol levels (hypercholesterolemia) have also been treated by interrupting the enterohepatic circulation of bile acids. Because a major use of cholesterol by the liver is in the production of bile acids, cholesterol sequestrants, such as cholestyramine resin, which blocks the reabsorption of bile acids, can reduce serum levels of cholesterol. As a result of blocking reabsorption, feedback regulation normally exerted by bile acids is reduced, and conversion of cholesterol to bile acids is enhanced in order to maintain an adequate bile acid pool.

What step commits cholesterol to bile acids synthesis?

The 7α-hydroxylation of cholesterol catalyzed by 7α-hydroxylase, a microsomal enzyme, commits cholesterol to the biosynthesis of bile acids. This activity is induced by dietary cholesterol and suppressed by bile acids via activation of protein kinase C. The phosphorylated form of 7α-hydroxylase is the active form (Figure 12.7).

What are the products derived from the bile acid pathway?

The pathway of bile acid biosynthesis divides into two pathways, one leading to cholyl-CoA and the other to chenodeoxycholyl-CoA. Cholyl-CoA has an extra α-OH group at position 12; other than this, the pathways involve similar hydroxylation reactions and shortening of the side chains to give typical bile acid structures of α-OH groups on positions 3 and 7 and full saturation of the steroid rings. These primary bile acids enter the bile as glycine and taurine conjugates and are probably in a salt form. In the intestine, the bile acids are subjected to further changes by intestinal bacteria that deconjugate and dehydroxylate at the 7α-OH position, producing secondary bile acids such as deoxycholic acid and lithocholic acid.

How much of the bile acids are reabsorbed in the intestine?

About 98% to 99% of the primary and secondary bile acids are absorbed almost exclusively in the ileum, returning to the liver by way of the portal circulation. The exception is lithocholic acid, which is not reabsorbed to any significant extent because of its insolubility. Although the amount of bile acid lost is only a very small fraction, this loss represents the major pathway for elimination of cholesterol. The amount that is lost in the feces is compensated for by synthesis of cholesterol in the liver so that the pool of bile acids is maintained.

Is there a storage form of cholesterol?

Cholesterol esters, in which the hydroxyl group of the cholesterol is esterified with a long-chain fatty acid, are used as an intracellular storage form of cholesterol. For example, cholesterol esters are abundant in steroid hormone-producing tissues, particularly the adrenal cortex, where they form lipid droplets in the cytoplasm. Cholesterol esters are also prominent in the plasma lipoproteins, where approximately 70% of the cholesterol is esterified.

What are lipoproteins?

The classes of lipoproteins are based on their composition. For example, the major triglyceride-rich lipoprotein is very LDL (VLDL), and approximately 70% of the total cholesterol is in LDL. Therefore, elevations of the plasma triglyceride level usually are caused by increased VLDL, whereas an elevated cholesterol level most often reflects an increase in the LDL fraction. The apolipoprotein contents of the different lipoproteins are also characteristic as well as their densities, which reflect the relative composition of lipid to protein in these particles.

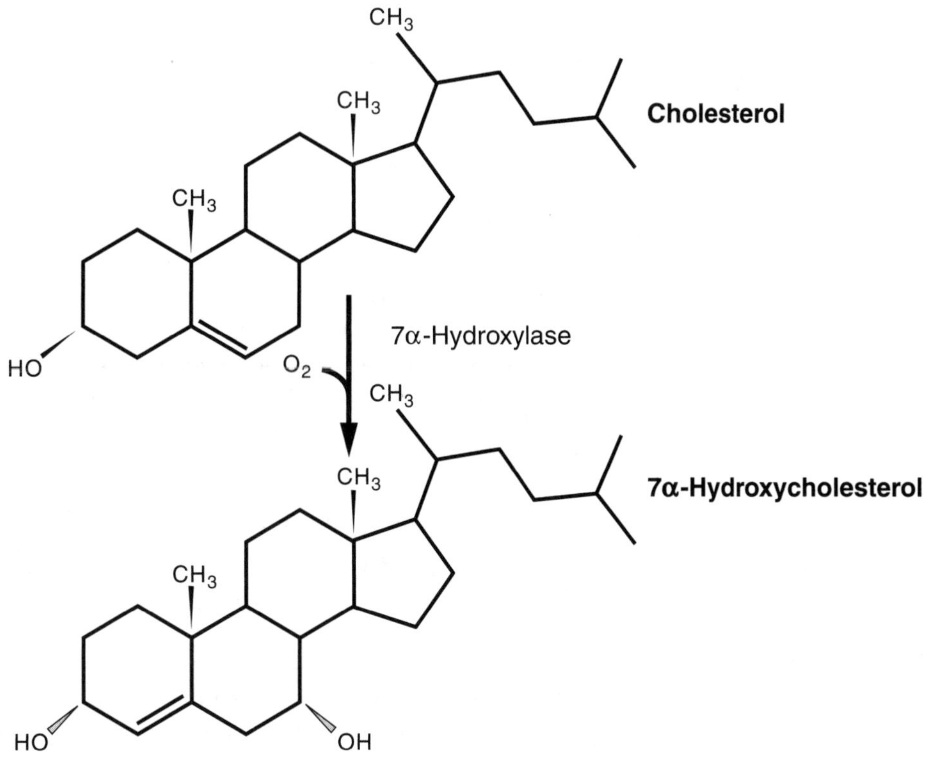

Figure 12.7 The 7α-hydroxylase reaction. The committed and regulated step in bile acid synthesis catalyzed by 7α-hydroxylase. The biosynthesis of bile acids represents the major metabolic fate of cholesterol in the human body.

Does a deficiency in the ability to assemble lipoproteins tell us something about the importance of their function?

There are rare incidences in which an inherited disease leads to a deficiency in an individual's ability to produce lipoproteins. For example, abetalipoproteinemia, a recessively inherited disease, is caused by the absence of triglyceride transfer protein located in the endoplasmic reticulum. As a result, liver and intestine are unable to assemble and secrete the triglyceride-rich apoB-100–containing lipoproteins. LDL, VLDL, and chylomicrons are essentially absent. However, the delivery of liver cholesterol to extrahepatic tissue, although compromised, can still occur via high-density lipoprotein (HDL) particles containing apoE, which is recognized by extrahepatic "LDL" receptors. Nevertheless, these individuals suffer from severe fat malabsorption and steatorrhea, which is accompanied by severe deficiencies of the fat-soluble vitamins. Triglycerides build up in the intestinal mucosa and liver.

How is cholesterol transported between tissues?

The liver releases cholesterol and cholesterol esters as constituents of VLDL, together with triglycerides and phospholipids. VLDL is processed to LDL, which is taken up by both the liver and extrahepatic tissues. Unlike the capillary endothelium of the intestinal mucosa, the sinusoidal endothelium of the liver is fenestrated, which allows passage of the lipoproteins into the sinusoidal blood. Newly synthesized VLDL contains apoB-100 and small amounts of apoE and the C-apolipoproteins. VLDL later acquires more apoE and C-apolipoproteins by transfer from HDL. Like chylomicrons, VLDL is metabolized initially by lipoprotein lipase. VLDL triglycerides are hydrolyzed, and the free fatty acids are taken up by tissues. Triglyceride hydrolysis is accompanied by transfer of the C-apolipoproteins to HDL, and approximately half of the VLDL remnants, especially the larger particles containing multiple copies of apoE, are taken up into the liver. In contrast, the smaller remnants appear as intermediate density lipoprotein (IDL) and are eventually remodeled to LDL.

How are the larger VLDL remnants (IDL) processed to form LDL?

The processing of IDL requires the hydrolysis of excess triglyceride and phospholipid by hepatic lipase (HL), as well as the transfer of excess apolipoproteins to HDL. Hepatic lipase is found on cell surfaces in the space of Disse, where it is anchored to heparan sulfate proteoglycans. Unlike lipoprotein lipase, hepatic lipase is not activated by apoC-II and does not attack triglycerides in chylomicrons or VLDL. LDL now has a well-defined structure possessing only apoB-100 and a high proportion of cholesterol and cholesterol esters as its lipid component.

What happens to the cholesterol in LDL?

The metabolism of LDL is slow relative to VLDL. LDL circulates for an average of 3 days and is eventually taken up by receptor mediated endocytosis, which is initiated by receptor binding apoB-100. Approximately two thirds of the LDL is taken up by the liver and one third by extrahepatic tissue. In extrahepatic tissue, LDL uptake represents the major source of exogenous cholesterol.

Why is the LDL carried cholesterol considered the "bad cholesterol?"

Not all LDL is cleared by the LDL receptor. Macrophage and some endothelial cells possess alternative lipoprotein receptors, collectively known as the scavenger receptors. These receptors have a lower affinity for LDL, which means that their contribution to LDL metabolism is greatest when the plasma LDL concentration is high. The scavenger receptors have even a greater affinity for LDL that has been chemically modified by acetylation or oxidizing agents. In fact, oxidized LDL may be a prime cause of atherosclerosis because it is taken up via scavenger receptors of macrophages and converted to foam cells that form fatty streaks, considered to be the precursors of atheromas. LDL particles enter the intima by transcytosis through the endothelium and are then endocytosed by tissue macrophage through their scavenger receptors. The uptake of LDL cholesterol has to be balanced by the transfer of excess cholesterol from the macrophage to HDL for transport back to the liver. Therefore, a foam cell and a fatty streak develop only if the amount of cholesterol acquired from LDL exceeds the amount released to HDL.

What particular disease emphasizes the cardiovascular risks associated with elevated levels of LDL?

Familial hypercholesterolemia is an autosomal dominant inherited disease caused by a deficiency of LDL receptors in liver and extrahepatic tissue. As a result, in the heterozygote, LDL accumulates to approximately twice its normal concentration, and total plasma cholesterol is in the 250- to 500-mg/dL range. Interestingly, cells do not suffer from a lack of cholesterol because they can compensate by increased endogenous synthesis. In affected heterozygotes, functional LDL receptors are reduced by approximately 50%, and these individuals may present with their first myocardial infarction by the age of 30 years.

Why is HDL considered a "good form" of cholesterol in the blood?

There is an inverse relationship between HDL concentrations and coronary heart disease, and the ratio between LDL and HDL may be the most predictive of risk. This relationship may reflect the fact that LDL is transporting cholesterol to the tissues and HDL acts as the scavenger of cholesterol in returning it back to the liver.

What is lipoprotein(a) [Lp(a)]?

There is a form of LDL in which an unusual glycoprotein, apolipoprotein(a), is associated through a disulfide bridge to the apoB-100 molecule. Elevated concentrations of this LDL form, Lp(a), have been shown to be an independent risk factor for atherosclerosis.

Steroid Metabolism

What tissue is responsible for the synthesis of most of the steroid hormones?

The adult adrenal cortex has three distinct layers or zones called the zona glomerulosa, zona fasciculata, and zona reticularis. Each layer is responsible for the synthesis of specific types of steroid hormones. The zona glomerulosa is associated with the production of mineralocorticoids. The zona fasciculata and zona reticularis produce glucocorticoids and androgens (Figure 13.1).

Steroid hormones are synthesized from what compound?

The adrenal steroid hormones are synthesized from cholesterol.

What is the regulated step in steroid biosynthesis?

The committed and regulated step in the pathway for steroid synthesis is catalyzed by an enzyme called P_{450} side-chain cleavage enzyme. All mammalian steroid hormones are formed from cholesterol via pregnenolone through a series of reactions that occur in either the mitochondria or endoplasmic reticulum of adrenal cells (Figure 13.2).

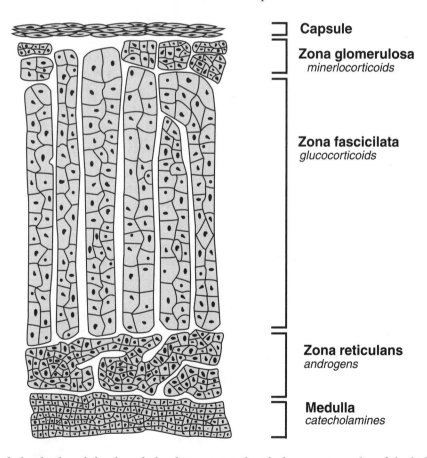

Capsule

Zona glomerulosa
minerlocorticoids

Zona fascicilata
glucocorticoids

Zona reticulans
androgens

Medulla
catecholamines

Figure 13.1   The adrenal gland. The adult adrenal gland is associated with the superior poles of the kidneys. Each gland consists of a yellowish outer cortex that has three distinct layers or zones called the zona glomerulosa, zona fasciculata, and zona reticularis. Each layer is responsible for the synthesis of specific types of steroid hormones. Cells of the zona glomerulosa produce the mineralocorticoid aldosterone. The zona fasciculata and zona reticularis produce glucocorticoids and androgens. The functional distinctions between these latter two areas are not precise and they appear as a functional unit.

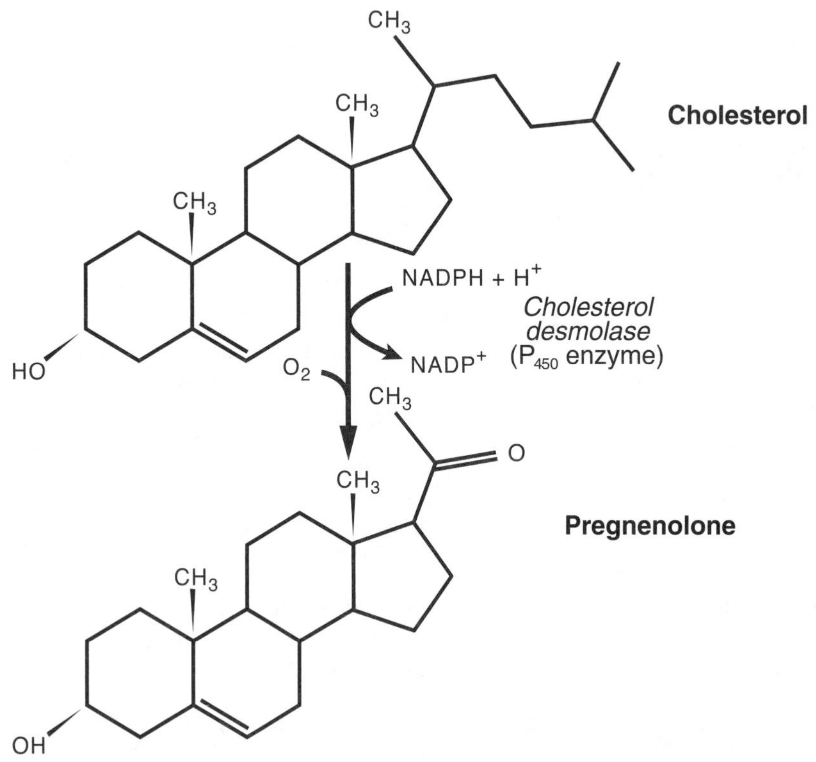

Figure 13.2 The cholesterol desmolase reaction. Cholesterol desmolase is the committed and regulated step in steroid biosynthesis, and is also known as cholesterol side-chain cleavage enzyme, a P_{450} enzyme. This P_{450} enzyme hydrolyzes the side chain of cholesterol at C_{20} and C_{22} to yield prenenolone, the precursor to all other steroid hormones.

What regulates this pathway?

After stimulation of the adrenals by ACTH that upregulates P_{450} side-chain cleavage enzyme, an esterase is also activated, releasing cholesterol that is transported into the mitochondrion where P_{450}scc converts cholesterol to pregnenolone.

After pregnenolone is formed, what types of activities are involved in synthesizing the various steroid hormones?

Essential activities include hydroxylases, dehydrogenases, and isomerases.

How is the synthesis of a particular steroid confined to a region of the adrenal cortex?

For example, two enzymes involved in aldosterone synthesis, 18-hydroxylase and 18-hydroxysteroid dehydrogenase, are found only in the zona glomerulosa. Thus, mineralocorticoid synthesis occurs only in this region of the adrenal cortex.

Where does glucocorticoid synthesis occur in the adrenal cortex?

Glucocorticoid synthesis requires the action of 17α-hydroxylase acting on either pregnenolone or progesterone. Through repeated shuttling of substrates into and out of the mitochondria of the fasciculata and reticularis cells, the glucocorticoids such as cortisol are synthesized.

What is the major androgen produced by the adrenal cortex?

The major androgen or androgen precursor is dehydroepiandrosterone. Although most of the 17-hydroxypregnenolone is directed to glucocorticoid synthesis in the adrenals, a small fraction has the two-carbon side chain removed by the action of a 17,20 lyase. This enzyme is found not only in the adrenal, but also in the gonads. Importantly, if glucocorticoid biosynthesis is blocked by a deficiency in one of the hydroxylases, then androgen production will necessarily increase.

How is the glucocorticoid synthesis regulated?

Cortisol synthesis is regulated by the diurnal rhythm of ACTH release from the pituitary. Cortisol in the blood inhibits production of corticoid-releasing hormone in the hypothalamus. This, in turn, prevents the release of ACTH by the pituitary.

How are glucocorticoid hormones carried in the serum?

Cortisol and its derivatives at normal levels of production are essentially found bound in circulation to a plasma-binding protein (α-globulin) called transcortin or corticosteroid-binding globulin. The avidity of binding, in part, determines the biologic half-life of the glucocorticoids.

Are the mineralocorticoids also carried by a specific plasma transport protein?

Aldosterone, the most potent of the natural mineralocorticoids, does not have a specific plasma transport protein, and as a result, it is cleared rapidly from the plasma by the liver.

How is mineralocorticoid synthesis regulated?

The primary regulator of mineralocorticoid synthesis is the renin–angiotensin system and potassium. ACTH, sodium, and neural mechanisms are also involved.

How do steroid hormones mediate a tissue-specific response?

Steroid hormones and thyroid hormone regulate a variety of processes involved in development, differentiation, growth, reproduction, and adaptation to environment. Steroid hormones regulate these processes in tissues that express the specific steroid receptor. Unlike peptide hormones, the receptors for steroid hormones are intracellular, and they mediate their response by interacting with specific DNA sequences called hormone-responsive elements in the nucleus of responsive cells.

What are the common features of steroid (and thyroid) hormone receptors?

The steroid and thyroid receptors belong to a gene superfamily that shares the following common features. Each receptor shares a region called a DNA binding domain that allows specific interaction of the receptor with its specific hormone responsive element (DNA sequence). Each receptor also contains a ligand binding domain that allows specific binding of the appropriate steroid hormone. There is also a region that all receptors share that allows for transactivation. This site of allows for specific protein–protein interaction with a transcriptional preinitiation complex. This interaction may promote or enhance the initiation of transcription of a gene under the control of a steroid hormone responsive element (Figure 13.3).

What causes Cushing's syndrome?

Cushing's syndrome is caused by an excessive production of glucocorticoids. This can be caused by the pharmacologic use of steroids, but it also may result from an ACTH-secreting pituitary adenoma, or from adrenal adenomas or carcinomas. It may even result from the ectopic production of ACTH by a neoplasm.

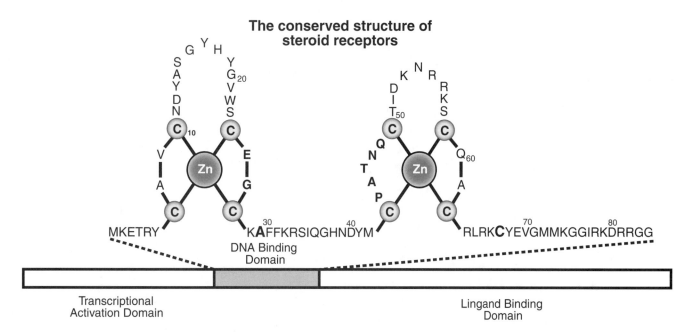

Figure 13.3 The conserved structure of steroid receptors. Steroid hormone receptors are located in the cell and share common and conserved features. One of the most notable is the zinc finger motif, which makes up the DNA binding domain and promotes interaction with the hormone responsive element in DNA.

What are the consequences of Cushing's syndrome?

Patients typically lose the diurnal pattern of ACTH and cortisol secretion. They have hyperglycemia or glucose intolerance because of accelerated gluconeogenesis. They also have related protein catabolic effects that result in thinning of the skin, muscle wasting, osteoporosis, extensive lymphoid tissue involution, and a general negative nitrogen balance. They also present with a characteristic redistribution of fat, truncal obesity, and the typical "buffalo hump."

How does a deficiency of a steroidogenic enzyme lead to congenital adrenal hyperplasia?

The important feature of this disease results from the deficiency of cortisol production with ACTH overproduction and adrenal hyperplasia. Two types of 21-hydroxylase deficiency account for more than 90% of cases of congenital adrenal hyperplasia. Partial deficiency of 21-hydroxylase leads to simple virilizing. Complete deficiency leads also to "salt wasting." Deficiency in 11β-hydroxylase accounts for the remainder of these cases.

Vitamin D and Calcitriol Metabolism

What is the primary source of vitamin D?

Small amounts of vitamin D occur in food such as fish liver oil, egg yolk, and supplemented milk. However, most of vitamin D that is made available for calcitriol synthesis is produced in the malpighian layer of the epidermis from 7-dehydrocholesterol in an ultraviolet light-induced, nonenzymatic photolysis reaction (Figure 13.4).

What is calcitriol?

Calcitriol (1,25 cholecalciferol) is a hormone, derived from vitamin D, that activates biologic processes in a manner that is similar to that employed by the steroid hormones. In other words, calcitriol mediates its response by binding to an intracellular receptor. The receptor ligand interacts with DNA at specific response elements to affect transcriptional regulation.

How is vitamin D (cholecalciferol) converted to calcitriol (1,25 cholecalciferol)?

Vitamin D_3 (cholecalciferol) is transported to the liver by a specific binding protein (vitamin D binding protein). In the liver, vitamin D_3 is converted to 25-hydroxycholecalciferol by a NADPH-dependent cytochrome P_{450} reductase and a cytochrome P_{450}. The 25-hydroxycholecalciferol enters circulation and is transported to the kidney. In the kidney, the 25-hydroxycholecalciferol is converted by a mitochondrial enzyme, 1α-hydroxylase, to 1,25-hydroxycholecalciferol or calcitriol (Figure 13.5).

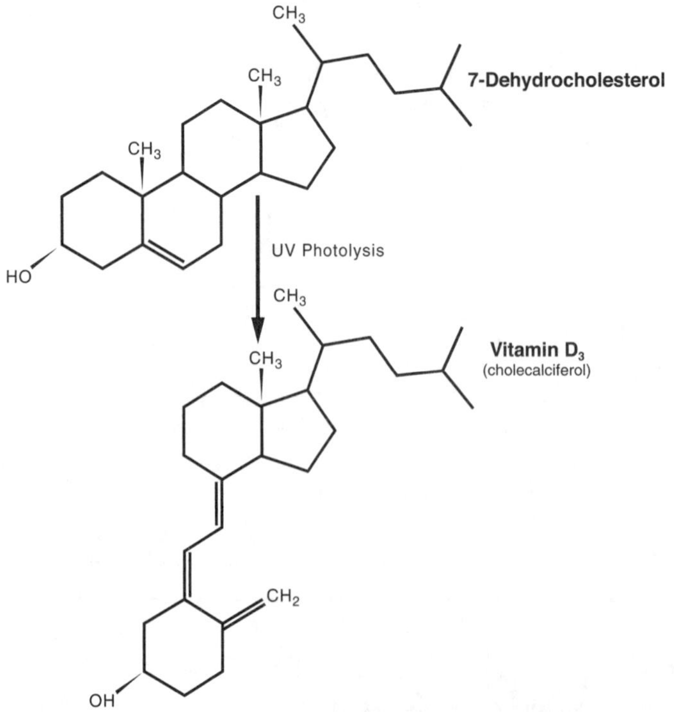

Figure 13.4 Light-induced, nonenzymatic photolysis of 7-dehydrocholesterol to vitamin D_3. 7-Dehydrocholesterol is an intermediate in the pathway of cholesterol biosynthesis and is converted in skin by UV rays (285–310 nm) to vitamin D_3.

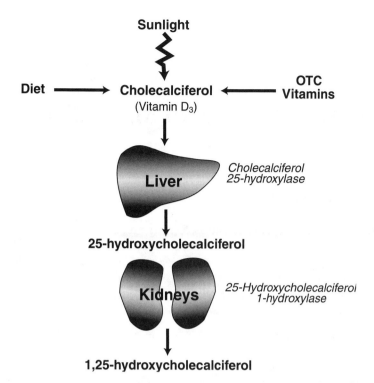

Figure 13.5 Synthesis of 1, 25-cholecalciferol, the hormonally active form of vitamin D_3. Cholecalciferol undergoes two successive hydroxylations. The first, at carbon 25 involves a microsomal enzyme system in the liver. 25-Hydroxycholecalciferol is then transported to the kidney, where a mitochondrial enzyme hydroxylates 25-cholecalciferol at carbon 1 to yield 1, 25-dihydroxycholecalciferol.

How is the synthesis of calcitriol regulated?

Calcitriol synthesis is controlled by feedback regulation. Low-calcium diets and hypocalcemia result in marked increases in 1α-hydroxylase activity. This effect requires parathyroid hormone, which is also released in response to low serum calcium. Calcitriol also regulates its own 1α-hydroxylase activity. High levels of calcitriol inhibit 1α-hydroxylase and stimulate 24-hydroxylase activity. The 24-hydroxylation appears to lead to inactive 24,25 cholecalciferol.

What is the biological role of calcitriol?

Calcitriol promotes the translocation of calcium against the concentration gradient, which exists across the intestinal cell membrane. Calcitriol also plays a significant role in controlling extracellular Ca^{2+} to ensure proper concentration of calcium and phosphate for deposition as hydroxyapatite crystals onto the collagen fibrils in bone.

How does a deficiency in calcitriol manifest itself in humans?

Rickets is most commonly caused by vitamin D deficiency. There are two types of vitamin D–dependent rickets. Type I is an inherited autosomal recessive trait characterized by a defect in 1α-hydroxylase. Type II is an autosomal recessive disorder where the defect is in the calcitriol receptor.

14 Amino Acid Metabolism

Amino Acids, Proteins, and Nutrition

What types of compounds other than proteins do amino acids serve as precursors?

Amino acids not only serve as the monomeric units of proteins, but also as precursors for other amino acids, and other essential groups of diverse compounds. These include the porphyrins, phospholipids, catecholamines and other hormones and neurotransmitters, the nicotinic acid of the pyridine nucleotides, purines, pyrimidines, the pigment melanin, the phosphagen creatine, carnitine, the polyamines spermine and spermidine, and probably many other compounds, some of whose functions are not yet known (Figure 14.1).

Does dietary protein serve as a sufficient source of amino acids for the synthesis of protein and the various other essential metabolites derived from amino acids?

The composition of dietary protein does not necessarily correspond to the individual's needs for the synthesis of protein and numerous other essential metabolites derived from amino acids. Therefore, metabolic pathways exist to derive energy from amino acids and to synthesize certain amino acids from other dietary constituents such as carbohydrates and fats, which can supply carbon, hydrogen, and oxygen atoms. However, the nitrogen is contributed almost entirely by ingested amino acids.

What is the basis for listing an amino acid as essential?

An amino acid that is needed for protein that cannot be produced by human metabolism is the basis for listing those amino acids as essential. With one exception, those amino acids listed as essential cannot be synthesized by the hu-

Metabolic Relationships of Amino Acids

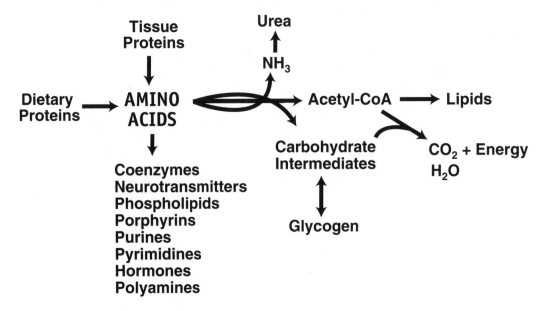

Figure 14.1 Amino Acid Metabolism

man body because our cells lack the necessary biosynthetic enzymes. In adults, the *de novo* synthesis of arginine apparently can meet normal requirements, but in children, normal growth requires an exogenous supply of this amino acid.

Under what circumstance can a nonessential amino acid become essential?

In some enzyme deficiency diseases, requirements for amino acids (i.e., tyrosine and cysteine) must be met by the diet because the deficiency prevents their *de novo* synthesis.

What does the term *nitrogen balance* mean?

The amount of nitrogen in each individual is regulated to stay relatively constant except during growth, when the amount must be increased in proportion to growth. There is no storage form for nitrogen reserves; only a small part of this nitrogen exists in the form of free amino acids or other compounds that can be used for synthesis of amino acids. Therefore, for maintenance of optimum body structure, an adequate supply of amino acids must be consumed frequently. If for any reason the supply of protein in the diet is insufficient, the need to synthesize specific proteins for vital physiologic functions results in a redistribution of amino acids among proteins.

How does the body compensate when the supply of protein in the diet is insufficient to maintain synthesis of specific proteins for vital physiologic functions?

For example, hemoglobin is degraded to the extent of almost 1% a day as red blood cells die, and under normal circumstances, the degradation is balanced by resynthesis. In a deficiency of amino acids, relatively less hemoglobin is synthesized because a degree of anemia is more tolerable than a deficiency of certain other proteins. In addition, the protein of muscle tissue can be degraded to supply a source of amino acids to maintain vital physiologic function.

Under normal dietary conditions, what happens to the bulk of the nitrogen derived from amino acids as a consequence of protein digestion?

Because normal diets in our country typically include a large excess of amino acids over the amount needed for synthesis of proteins or other cell constituents, most excess amino acids are degraded to products that are either oxidized for energy or stored as fat and glycogen. In each case, the nitrogen from the amino acid backbone is liberated as ammonia. Some of this ammonia is reused in the synthesis of amino acids. Some is used in other biosynthetic reactions. Some is excreted in the urine, but the largest part is converted to a compound called urea, which is excreted by the kidneys.

Protein Digestion

What role does the stomach play in the digestion of dietary protein?

In humans, the degradation of ingested proteins into their constituent amino acids occurs in the gastrointestinal tract. Entry of protein into the stomach stimulates the gastric mucosa to secrete the hormone gastrin, which in turn stimulates the secretion of hydrochloric acid by the parietal cells of the gastric glands and pepsinogen by the chief cells.

What role does the acidification of the stomach play in the digestion of protein?

Globular proteins denature at low pH, rendering their internal peptide bonds more accessible to enzymatic hydrolysis. Pepsinogen (40 kDa), and inactive precursor or zymogen, is converted into active pepsin in the gastric juice by the enzymatic action of pepsin itself. Thus, after favorable pH conditions are reached, pepsinogen is converted to pepsin by autoactivation and subsequently autocatalysis at an exponential rate. Pepsin hydrolyzes ingested proteins at peptide bonds on the amino-terminal side of the aromatic amino acid residues Tyr, Phe, and Trp. The major products of pepsin action are large peptide fragments and some free amino acids. The importance of gastric protein digestion does not lie so much in its contribution to the breakdown of ingested macromolecules, but rather in the generation of peptides and amino acids that act as stimulants for cholecystokinin release in the duodenum. The gastric peptides therefore are instrumental in the initiation of the pancreatic phase of protein digestion.

What role does the acidic stomach contents play in the further digestion of protein in the intestine?

As the acidic stomach contents pass into the small intestine, the low pH triggers the secretion of the hormone secretin from the intestinal cells into the blood. Secretin stimulates the pancreas to secrete bicarbonate into the small intestine, which neutralizes the gastric HCl, increasing the pH abruptly to about pH 7.

What signal triggers the further digestion of the partially degraded protein as it enters the intestine?

The entry of gastric peptides into the upper part of the intestine (duodenum) releases the hormone cholecystokinin from intestinal cells, which stimulates secretion of several pancreatic enzymes, whose activity optima occur at pH 7 to 8. Three of these enzymes—trypsin, chymotrypsin, and carboxypeptidase—are made by the exocrine cells of the pancreas as their respective enzymatically inactive zymogens.

Why are digestive proteases synthesized as zymogens?

Synthesis of digestive proteases as inactive precursors protects the exocrine cells from destructive proteolytic attack. Free trypsin can activate not only trypsinogen, but also three other digestive zymogens.

How does the pancreas protect itself from the possible autoactivation of trypsinogen to trypsin?

The pancreas also protects itself against self-digestion in another way—by making a specific inhibitor called pancreatic trypsin inhibitor. Trypsin inhibitor, a protein itself, effectively prevents premature production of free proteolytic enzymes within the pancreatic cells.

How are the zymogens activated in the small intestine?

After trypsinogen enters the small intestine, it is converted into its active form by enteropeptidase (enterokinase), a specialized proteolytic enzyme secreted by intestinal cells. Trypsin then activates not only trypsinogen, but also three other digestive zymogens. Trypsin and chymotrypsin hydrolyze peptides that result from the action of pepsin in the stomach.

How does the digestion of the short peptides from trypsin and chymotrypsin digestion result in amino acids?

Degradation of the short peptides in the small intestine is completed by other peptidases. The first is carboxypeptidase, a zinc-containing enzyme, which removes successive carboxyl-terminal residues from very short peptides. The small intestine also secrets an aminopeptidase. This enzyme hydrolyzes successive amino-terminal residues from these short peptides.

Can all types of proteins be digested by this process?

In humans, most globular proteins from animal sources are almost completely hydrolyzed into amino acids. However, some fibrous proteins, such as keratin, are only partially hydrolyzed. Many proteins of plant foods, such as cereal grains, are incompletely digested because the protein part of grains of seeds is surrounded by indigestible cellulose husks.

How is acute pancreatitis related to protein digestion?

Acute pancreatitis is caused by obstruction of the normal pathway for secretion of pancreatic juice into the intestine. Under these conditions, the zymogens of the proteolytic enzymes are converted into their catalytically active forms prematurely inside the pancreatic tissue. As a result, these enzymes attack the pancreatic tissue itself, causing a painful and serious destruction of the organ, which can be fatal.

By what type of process are free amino acids taken up from the intestine into the blood?

Intestinal cells take up free amino acids and some dipeptides and tripeptides by special transport proteins. All of the peptides are hydrolyzed to free amino acids in enterocytes before their release into the hepatic portal vein. Free amino acids are translocated by the luminal epithelial cells of the intestines using a Na^+-dependent symport transport system similar to the Na^+-dependent glucose transporter. A least four different translocators have been identified: (1) one for neutral amino acids such as alanine, valine, and leucine; (2) one for basic amino acids, including lysine and arginine; (3) one for the acidic amino acids aspartate and glutamate; (4) one for the amino acids proline and glycine. The free amino acids enter the blood capillaries in the villi and are transported to the liver.

How does the Na^+-dependent transport pathway translocate amino acids?

The chemical mechanism for the symport movement of molecules using the Na^+ ion gradient involves a cooperative interaction of the Na^+ ion and the other molecule translocated on the protein. A conformational change of the protein occurs after association of the two ligands (Na^+ and amino acid), which moves them the necessary distance to bring them into contact with the cytosolic environment. The dissociation of the Na^+ ion from the transporter because of the low Na^+ ion concentration inside the cell leads to the return of the protein to its original conformation, a decrease in the affinity for the amino acid, and a release into the cytosol (Figure 14.2).

Does a defect in the ability to transport an amino acid across a membrane ever lead to a deficiency in that amino acid?

There are generally two causes of amino acid deficiency: (1) low dietary intake or (2) impaired absorption. One of the significant features of the amino acid transport system is that an overabundance of a similar amino acid in the diet may cause an apparent deficiency because of competitive inhibition in the transport process, particularly in the gut. Thus, ingestion of large quantities of one amino acid may prevent the absorption of another amino acid that uses the same transporter. There are also pathologic conditions caused by alteration in the transport systems for specific amino acids. For example, in Hartnup's disease there is a deficiency in the transport of neutral amino acids in

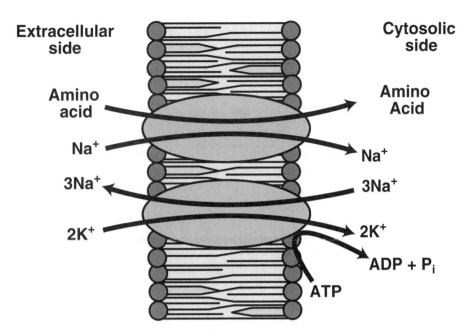

Figure 14.2 Na$^+$-dependent amino acid transport system. Most amino acids are cotransported from the intestinal lumen with Na$^+$ across the apical plasma membrane into mucosal epithelial cells. The amino acid moves through the cell to the basal surface, where it passes into the blood.

the epithelial cells of the intestine and renal tubules. In Cystinuria, the renal absorption of cystine and the basic amino acids lysine and arginine is abnormal, resulting in formation of cystine kidney stones.

Nitrogen Metabolism and Ammonia Detoxification

What is the major source of ammonia nitrogen?

Amino acids derived from dietary proteins are the source of most amino groups. Most amino acids are metabolized in the liver. Some of the ammonia that is generated is recycled and used in a variety of biosynthetic processes; the excess is either excreted or converted to urea for excretion.

Under what metabolic circumstances do amino acids undergo oxidative degradation?

Amino acids can undergo oxidative degradation in three different metabolic circumstances: (1) During the normal synthesis and degradation of cellular proteins, some of the amino acids are released during protein breakdown and undergo oxidative degradation if they are not needed for new protein synthesis. (2) When a diet is rich in protein, amino acids are ingested in excess of the body's need for protein synthesis; thus, the surplus may be catabolized because amino acids cannot be stored. (3) During starvation or in diabetes mellitus, when carbohydrates are either unavailable or not properly used, body proteins are mobilized as fuel source.

What is the essential first step that amino acids must undergo in oxidative degradation?

Amino acids lose their amino groups to form α-ketoacids. These carbon skeletons (α-ketoacids) of the amino acids provide three and four carbon units that can be converted to glucose, which in turn can fuel the functions of the brain, muscle, and other tissues.

What happens to the α-amino groups?

If α-amino groups of the 20 L-amino acids are not reused for synthesis of new amino acids or other nitrogenous products, these amino groups are channeled into a single excretory end product, urea.

What is the mechanism for the removal of the amino groups?

The removal of the α-amino groups is catalyzed by enzymes called aminotransferases or transaminases where the α-amino groups are transferred to the α-carbon atom of α-ketoglutarate, leaving behind the corresponding α-keto acid analogue of the amino acid. This is essentially the first step in the catabolism of most of the L-amino acids (Figure 14.3).

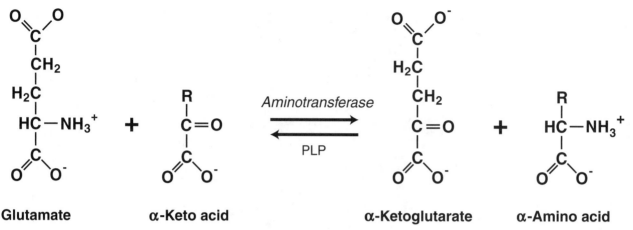

| Glutamate | α-Keto acid | | α-Ketoglutarate | α-Amino acid |

Figure 14.3 Aminotransferase reaction (transamination). The aminotransferase catalyzed reaction provides a major route for both synthesis and degradation of most amino acids. The enzyme requires the cofactor pyridoxal phosphate (PLP) that is derived from vitamin B_6. The reaction is highly reversible, thus the direction of the reaction is dependent on the relative concentration of the reaction constituents.

What is the overall significance of the transamination reaction?

There is no net deamination in such reactions because the α-ketoglutarate becomes aminated as the α-amino acid is deaminated, and the significance of this is that the effect of transamination reactions is to collect the amino groups from many different amino acids in the form of only one, namely L-glutamate. Glutamate channels amino groups either into biosynthetic pathways or into a final sequence of reactions by which nitrogenous waste products are formed and then excreted.

What role does vitamin B_6 (pyridoxine) play in the aminotransferase reaction?

All aminotransferases share a common prosthetic group and a common reaction mechanism. The prosthetic group is pyridoxal phosphate, the coenzyme formed from pyridoxine or vitamin B_6, that functions as an intermediate carrier of amino groups at the active site of aminotransferases.

Why are serum levels of certain aminotransferases monitored and used as a diagnostic tool?

The measurement of alanine aminotransferase (ALT, also known as GPT for glutamate-pyruvate transaminase) and aspartate aminotransferase (AST, also known as GOT for glutamate-oxaloacetate transaminase) levels in blood serum are used as an indicator of heart or liver damage because their levels in the blood increase significantly as a result of their release from the damage of these tissues.

What is the function of glutamate in ammonia utilization or detoxification?

Glutamate is transported from the cytosol to the mitochondria, where it undergoes oxidative deamination catalyzed by glutamate dehydrogenase. Glutamate dehydrogenase is present only in the matrix of the mitochondria and requires NAD^+ or $NADP^+$ as the acceptor of the reducing equivalents. The combined action of aminotransferases and glutamate dehydrogenase is referred to as transdeamination (Figure 14.4).

Why is the activity of glutamate dehydrogenase regulated?

As might be expected from its central role in amino acid metabolism, glutamate dehydrogenase is a complex allosteric enzyme. It is influenced by the positive modulator ADP and by the negative modulator GTP, a product of the succinyl-CoA synthetase reaction in the citric acid cycle. When ADP levels increase, glutamate dehydrogenase activity increases, making α-ketoglutarate available for the TCA cycle and releasing NH_4^+ for excretion. When ATP levels are high, thus GTP high, in the mitochondria as a result of high TCA cycle activity, oxidative deamination of glutamate is inhibited. In addition, this reaction will be greatly influenced by the levels of either glutamate or NH_4^+ so that the direction of the reaction catalyzed by glutamate dehydrogenase depends on the level of available substrate. This, more than the energy state of the cell, most likely influences the direction of the reaction catalyzed by the enzyme.

What role does glutamine play in ammonia metabolism?

In many tissues, including the brain, ammonia is enzymatically combined with glutamate to yield glutamine by the action of the enzyme called glutamine synthetase. This reaction occurs in two steps and requires ATP. Glutamine is a nontoxic, neutral compound that can readily pass through cell membranes, whereas glutamate, which bears a net negative charge, cannot. L-glutamine is critical for carrying amino groups in transport to the liver. Glutamine is a major transport form of ammonia and is normally present in blood in much higher concentrations than other amino acids (Figure 14.5).

Glutamate dehydrogenase

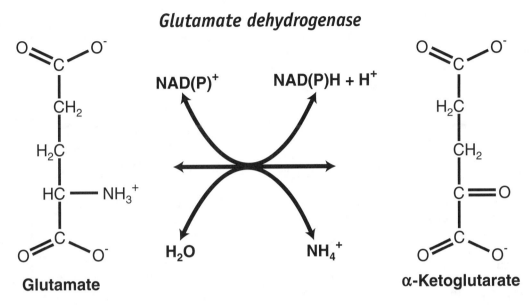

Figure 14.4 Glutamate dehydrogenase reaction. Glutamate dehydrogenase directs an important intersection of carbon and nitrogen metabolism. Amino groups from many of the α-amino acids are collected in the liver in the form of glutamate. In hepatocytes, glutamate is transported from the cytosol into the mitochondria, where glutamate dehydrogenase catalyzes the removal of the amino group in the form of NH_4^+, producing α-ketoglutarate. The NH_4^+ can be used directly in the synthesis of urea.

Glutamine synthetase

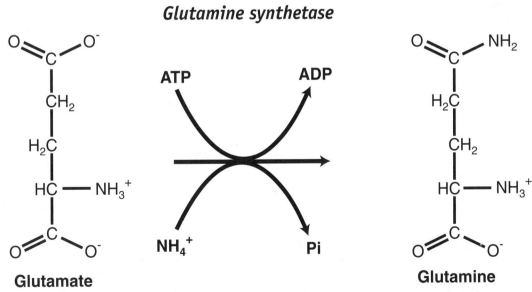

Figure 14.5 Glutamine synthetase reaction. Free ammonia produced in tissues is combined with glutamate to yield glutamine by the action of glutamine synthetase. The reaction requires ATP. Glutamine is a nontoxic transport form of ammonia and is normally found in blood in much higher concentrations than other amino acids. Glutamine also serves as a source of amino groups in a variety of biosynthetic reactions, which should be particularly noted in purine and pyrimidine *de novo* biosynthesis.

How does glutamine provide ammonia for urea formation in the liver?

In the mitochondria of the liver, glutaminase converts glutamine to glutamate and NH_4^+. The glutamate generated from glutamine is then further oxidized by glutamate dehydrogenase to α-ketoglutarate and NH_4^+ so as ultimately to deliver two equivalents of NH_4^+ for urea synthesis (Figure 14.6).

In addition to acting as a major carrier to transport ammonia to the liver, does glutamine serve any other function as an amino donor?

In addition to its role in the transport of amino groups, glutamine serves as a source of amino groups in over a dozen biosynthetic reactions. The enzymes catalyzing these reactions are called glutamine amidotransferases. The NH_3 produced by these enzymes remains at the active site and reacts with the second substrate to form the aminated product. The covalent intermediate is hydrolyzed to form the free enzyme and glutamate. Glutamine amidotransferases are particularly important in purine and pyrimidine metabolism.

PRPP amidotransferase
Example: Glutamine + PRPP → amidoPRPP + glutamate

In addition to glutamine, are there any other amino acids used to transport tissue ammonia to the liver from extra hepatic tissue?

In muscle, excess amino groups are generally transferred to pyruvate to form alanine. Alanine plays a special role in transporting amino groups from muscle to the liver in a nontoxic form via the glucose-alanine cycle.

Why is alanine used by muscle to transport amino groups to the liver?

In muscle and certain other tissues that degrade amino acids for fuel, amino groups are collected in glutamate by transamination. Glutamate may then be converted to glutamine for transport to the liver, or it may transfer its α-amino group to pyruvate, a readily available product of muscle glycolysis, via the reaction catalyzed by alanine aminotransferase. Alanine, with no net charge at pH near 7, passes into the blood and is carried to the liver. Vigorously contracting skeletal muscle operates anaerobically, producing not only ammonia from protein breakdown, but also large amounts of pyruvate from glycolysis. Both of these products must find their way to the liver where alanine can release its amino group by transamination to be converted to urea for excretion and the α-ketoacid of alanine, pyruvate, can be rebuilt via gluconeogenesis to glucose and returned to muscle (Figure 14.7)

Urea Cycle

Which step in urea synthesis is regulated?

The reaction catalyzed by carbamoyl phosphate synthetase I, where NH_4^+ is conjugated with HCO_3^- to yield carbamoyl phosphate, is the regulated step.

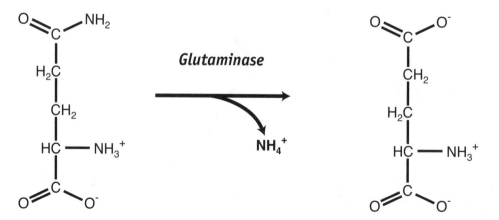

Figure 14.6 Glutaminase reaction. In humans, glutamine in excess of that required for biosynthesis is transported in the blood to the intestine, liver, and kidneys for processing. In these tissues, the amide nitrogen is released as ammonium ion in the mitochondria by the enzyme glutaminase.

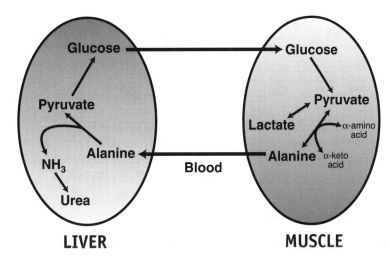

Figure 14.7 Glucose-alanine shuttle. In muscle and certain other tissues that degrade amino acids for fuel, amino groups are collected in the form of glutamate by transamination. Glutamate can be converted to glutamine for transport to the liver, or its amino group can be transferred to pyruvate by alanine aminotransferase. The alanine formed is transported in the blood to the liver, where alanine aminotransferase transfers the amino group from alanine to α-ketoglutarate, forming pyruvate and glutamate.

How is this carbamoyl phosphate synthetase I regulated?

The flux of nitrogen through the urea cycle varies with the composition of the diet. All five enzymes of the urea cycle are synthesized at higher levels during starvation or in animals with very high-protein diets. On a shorter time scale, however, allosteric regulation of carbamoyl synthetase I occurs by N-acetylglutamate. N-acetylglutamate synthetase is, in turn, activated by arginine, a urea cycle intermediate that accumulates when urea production is too low to accommodate the ammonia produced by amino acid catabolism.

What is the energetic cost of urea production?

The urea cycle is energetically expensive. The synthesis of one molecule of urea requires the equivalent of four high-energy phosphate bonds. It has been estimated that a human loses 15% of the energy of the amino acid from the urea derived.

Why does the hyperammonemia pose such a serious clinical problem?

The catabolic production of ammonia poses a serious problem because ammonia is very toxic, particularly to the central nervous system. The terminal stage of ammonia intoxication in humans is characterized by the onset of a comatose state and other effects on the brain. Although the molecular basis of ammonia toxicity in brain is not entirely understood, the major effects have been proposed to involve changes in cellular pH (alkalosis) and the depletion of certain TCA cycle intermediates (i.e., α-ketoglutarate).

How do elevated levels of ammonia affect cellular depletion of TCA cycle intermediates?

The removal of excess ammonia from the cell involves reductive amination of α-ketoglutarate to form glutamate by glutamine dehydrogenase and conversion of glutamate to glutamine synthetase. Both of these enzymes occur in high levels in the brain. The first reaction depletes the cellular NADH and α-ketoglutarate required for ATP production in the cell. The second reaction catalyzed by glutamine synthetase depletes the cell of ATP itself. Overall, NH_3 may interfere with the very high levels of ATP production required to maintain brain function.

Does depletion of glutamate to produce glutamine in brain tissue have any other potential consequences?

Depletion of glutamate in the glutamine synthetase reaction may have additional effects on the brain. Glutamate and the compound γ-aminobutyrate that is derived from glutamate are both important neurotransmitters. Sensitivity of the brain to ammonia may reflect a depletion of neurotransmitters.

Can people with genetic defects who have an impaired ability to convert ammonia to urea be treated?

The outcome for these people is very dependent on the nature of the defect. As an example, a child born with a deficiency in argininosuccinate synthetase must be diagnosed and treated as soon as possible. Argininosuccinate deficiency (citrullinemia) is an autosomal recessive disorder; the biochemical consequences include a failure to synthesize urea, resulting in the accumulation of nitrogen as ammonium, glutamine, and citrulline. Although the clinical course of untreated neonatal onset of this disease is remarkable in its consistency, there is considerable phenotypic variability. The most severe onset occurs in the neonatal period with a rapidly progressive encephalopathy and death. The disease may also be manifested later in childhood with episodic hyperammonemic encephalopathy and developmental delay. The therapeutic strategy for the long-term treatment of a patient with a deficiency in argininosuccinate synthetase exploits our understanding of ammonia metabolism and is based on alternative ways to excrete waste nitrogen. There is a restriction in nitrogen intake in the form of protein; arginine becomes an essential amino acid because it can no longer be synthesized and is also required to remove nitrogen in the formation of citrulline. Glutamine levels are diminished by exploiting alternative metabolic pathways. For example, ingestion of phenylbutyrate, which the liver converts to phenylacetate, can be exploited to lower glutamine levels because the liver subsequently conjugates phenylacetate with glutamine to facilitate excretion. Glutamine levels can be reduced and managed this way to prevent the build up of ammonia.

Catabolism and Precursor Functions of Amino Acids

What is the significance of the classification of amino acids as glucogenic or ketogenic?

Before detailed metabolism of amino acids had been established, the relationship of amino acids to fat and carbohydrate metabolism was established through physiologic experiments. The outcome of these experiments demonstrated that six amino acids were degraded to acetoacetyl-CoA and/or acetyl-CoA in the liver. Thus these six amino acids (tryptophan, phenylalanine, tyrosine, isoleucine, leucine, and lysine) were considered to be ketogenic. Whereas other amino acids were found to be converted into pyruvate, α-keytoglutarate, succinyl-CoA, fumarate and oxaloacetate, and as such were considered to be glucogenic. Importantly, however, the division between ketogenic and glucogenic amino acids was not distinct. For example, four amino acids (tryptophan, phenylalanine, tyrosine, and isoleucine) were shown to be both ketogenic and glucogenic (Figure 14.8).

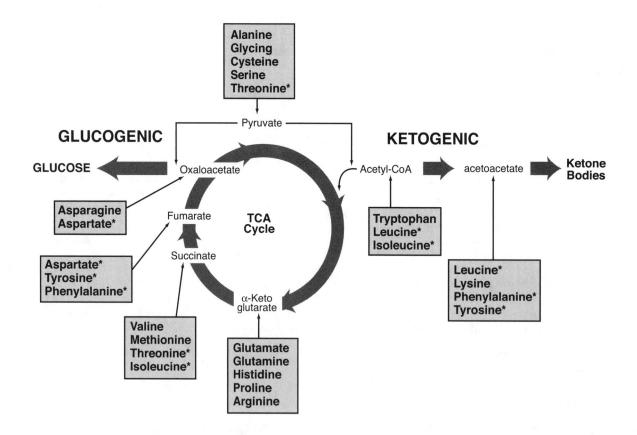

Figure 14.8 Metabolic fates of amino acid carbon skeletons. Catabolism of the 20 amino acids come together to form only six major products, all of which enter the TCA cycle. From here the carbon skeletons can be diverted to the formation of glucose (glucogenic) or acety-CoA (ketogenic). Five of the amino acids are both glucogenic and ketogenic. Only two of the amino acids, leucine and lysine, are exclusively ketogenic.

What is the medical significance of this classification of amino acids?

Some amino acids can be converted into pyruvate, particularly alanine; cysteine and serine are important in severe starvation and untreated diabetes mellitus. In addition, the ability of some amino acids to form ketones is evident in untreated diabetes mellitus, in which large amounts of ketones are produced in the liver, not only from fatty acids, but from ketogenic amino acids. In fact, degradation of leucine, an exclusively ketogenic amino acid that is very common in proteins, makes a substantial contribution to ketosis during starvation.

Why does the catabolism of branch-chain amino acids have metabolic significance to nonhepatic tissue?

Although much of the catabolism of amino acids occurs in the liver, the three branched-chain amino acids (leucine, isoleucine, and valine) are oxidized as fuels primarily in muscle, adipose, kidney, and brain tissue. These extrahepatic tissues contain a single aminotransferase not present in liver that acts on the three branched-chain amino acids to produce the corresponding α-keto acids.

How do the deaminated products of the branch-chain amino acids function as fuels in non-hepatic tissues?

The deaminated products of the branch-chain amino acids, as branch-chain α-ketoacids, mimic pyruvate, and as such are oxidized by an enzyme complex similiar to pyruvate dehydrogenase called the branch-chain α-ketoacid dehydrogenase. The branch-chain α-ketoacid dehydrogenase, like pyruvate dehydrogenase and α-ketoglutarate dehydrogenase, contains five co-factors: thyamine pyrophosphate, FAD, NAD, lipoate, and coenzyme A; and catalyzes a similiar series of reactions to produce branch-chain acyl-CoA detrivates, CO_2 and NADH, that feed electrons into the mitochondrial electron transport system. These branch chain acyl-CoAs are analogues of short chain fatty acyl-CoAs, and as such are further oxidized by specific dehydrogenases.

How might one treat a deficiency in branch-chain α-keto acid dehydrogenase?

This inherited defect, where the branch-chain α-keto acid dehydrogenase complex is deficient, is called maple syrup urine disease because of the characteristic odor imparted to the urine by the branch-chain α-keto acids. Treatment requires a rigid control over diet to limit intake of valine, isoleucine, and leucine to the minimum required to permit normal growth. This condition results in abnormal development of the brain, mental retardation, and death in early infancy.

One-Carbon Transfer Reactions

What is a one-carbon transfer reaction?

One-carbon transfers usually involve one of three cofactors: biotin, tetrahydrofolate, or S-adenosyl-L-methionine (SAM). These cofactors are used to transfer one-carbon groups in different oxidation states. The most oxidized of carbon, CO_2, is transferred by biotin. The remaining two cofactors are especially important in amino acid and nucleotide metabolism. Tetrahydrofolate is generally involved in transfer of one-carbon groups in the intermediate oxidation states, and S-adenosylmethionine in transfer of methyl groups, the most reduced state of carbon.

What is the major source of one-carbon units for tetrahydrofolate?

The major source of one-carbon units for tetrahydrofolate is the carbon removed in the conversion of serine to glycine, producing N^5,N^{10}-methylenetetrahydrofolate. This is a reversible reaction so that glycine may be converted to serine by enzymatic addition of a hydroxylmethyl group (Figure 14.9).

Does glycine have any other catabolic fates?

A second pathway for glycine predominates in animals and involves the oxidative cleavage of glycine into CO_2 and NH_4^+ and a methylene group of N^5,N^{10}-methylenetetrahydrofolate acid. This reaction is catalyzed by glycine cleavage enzyme (glycine synthase), and the significance of this pathway is exemplified by the severity of the disease that results from a deficiency of this enzyme known as *nonketotic hyperglycemia*. The severe mental retardation that is caused by a deficiency of the glycine cleavage enzyme may be accounted for based on the involvement of glycine as a major inhibitory neurotransmitter in the brain (Figure 14.10).

Why is the most pronounced effect of folate deficiency the inhibition of DNA synthesis?

Tetrahydrofolate derivatives are required in the synthesis of choline, serine, glycine, methionine, purines, and dTMP. Because adequate amounts of choline and amino acids can usually be obtained from the diet, the participation of folates in purine and dTMP synthesis appears to be the most metabolic significant. Thus, inhibition of DNA synthesis is due to the decreased availability of purines and dTMP.

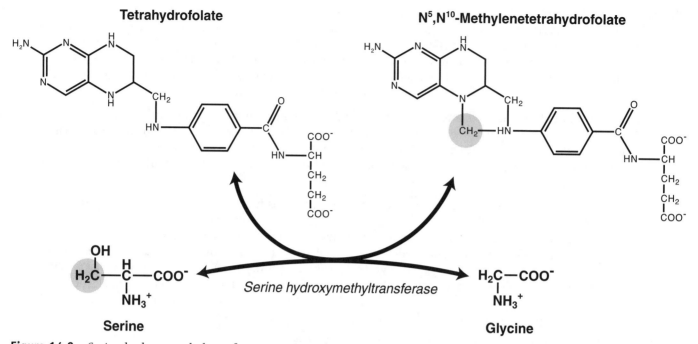

Figure 14.9 Serine hydroxymethyltransferase reaction. The conversion of serine to glycine by serine hydroxymethyltransferase to produce N^5,N^{10}-methylenetetrahydrofolate is a major source of one-carbon units in tetrahydrofolate metabolism.

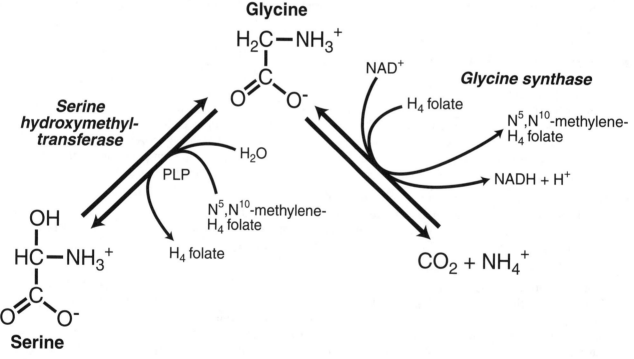

Figure 14.10 Glycine metabolism. In humans, glycine is degraded via two pathways. Glycine is converted to serine by the enzyme serine hydroxymethyltransferase, which requires the coenzymes tetrahydrofolate and pyridoxal phosphate. In the second pathway, glycine undergoes oxidative cleavage to yield CO_2, NH_4^+, and N^5,N^{10}-methylenetetrahydrofolate.

What significant aspect of histidine catabolism relates to folic acid deficiency?

    The enzyme that converts N-forminino-L-glutamate to glutamate in the histidine catabolic pathway requires tetrahydrafolic acid. Thus, increased levels of N-forminino-L-glutamate excretion in the urine have been used as an indicator of folic acid deficiency.

Why is the macrocytic anemia associated with megaloblastic changes in the bone marrow characteristic of folate deficiency?
A folate deficiency leads to inhibition of DNA synthesis due to a decreased availability of purines and dTMP. As a result, rapidly dividing cells arrest in S phase and develop a characteristic "megaloblastic" change in the size and shape of their nuclei. The block in DNA synthesis also slows the maturation of red blood cells, causing production of abnormally large "macrocytic" cells with fragile membranes.

Methyl Cycle and Succinyl-CoA Pathway

Why is SAM more commonly used for methyl group transfers in biosynthetic reactions?
Although tetrahydrofolate can carry a methyl group at N-5, the methyl group's transfer potential is insufficient for most biosynthetic reactions. SAM is a potent alkylating agent by virtue of its destabilizing sulfonium ion. The methyl group is subject to attack by nucleophiles and is about 1,000 times more reactive than the methyl group of N^5-methyltetrahydrofolate; thus, SAM is more commonly used for methyl group transfers.

What is the significance of the N^5-methyl derivative of tetrahydrofolate, which is the only form of tetrahydrofolate that is not interconvertible with other tetrahydrofolate forms, in a vitamin B_{12} deficiency?
The B_{12}-dependent homocysteine to methionine conversion by methionine synthase is the only major pathway by which N^5-methyltetrahydrofolate can return to the tetrahydrofolate pool. Thus, in B_{12} deficiency, there is a buildup of N^5-methyltetrahydrofolate and a deficiency of the tetrahydrofolate derivatives needed for purine and dTMP biosynthesis (Figure 14.11).

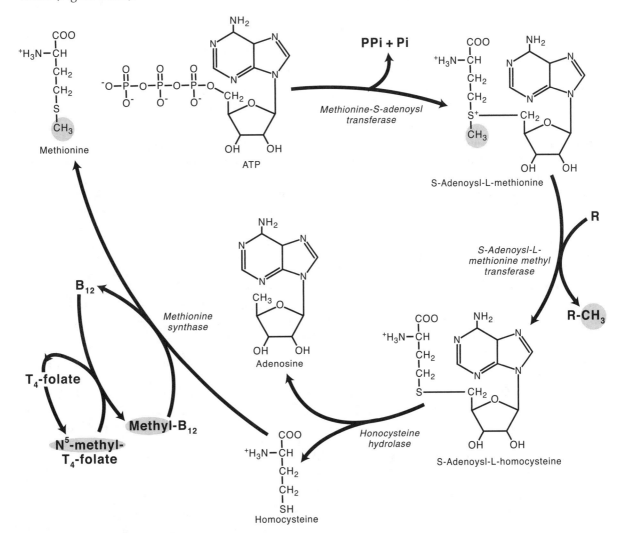

Figure 14.11   Methyl cycle. S-Adenosyl-L-methionine is the major donor for methyl group transfers. It is synthesized from ATP and methionine by the enzyme methionine adenosyl transferase. Methionine is regenerated in the methyl cycle from homocysteine and methyl-B_{12} by a reaction catalyzed by methionine synthase.

Is a deficiency in vitamin B_{12} uncommon?

Vitamin B_{12} is widespread in foods of animal origin, especially meats. Furthermore, the liver stores up to a 6-year supply of vitamin B_{12}. Thus, deficiencies of B_{12} are extremely rare. Most notably when they occur, they are observed in older people because of insufficient production of intrinsic factor and/or HCl in the stomach. Vitamin B_{12} can also be seen in patients with severe malabsorption diseases and in long-term vegetarians.

How do the major manifestations of B_{12} deficiency, hemopoietic and neurologic, relate to B_{12} metabolic functions?

The methyl derivative of B_{12} is required for the conversion of homocysteine to methionine. Since this is the only pathway by which N^5-methyl tetrahydrofolic acid can return to the tetrahydrofolate pool, most of the tissue tetrahydrofolic acid becomes "trapped" in the N^5-methyl form. This results in deficiency of precursors (especially dTTP) for DNA synthesis, which particularly effects hemopoietic tissue. It was once believed that a deficiency in B_{12} in the form of the 5-deoxyadenosine derivative required for the methylmalonyl-CoA mutase reaction resulted in neurological problems. However, more recent evidence suggests that the neuropathy associated with B_{12} deficiency may better relate to an induced imbalance in TNFα (tumor necrosis factor α) and EGF (epidermal growth factor) in the cerebrospinal fluid where appropriate expression of these two agents is critical to the normal maintenance and function of the CNS.

Why do homocysteine levels increase as a result of a B_{12} (cobalamin) or folic acid deficiency?

Methionine is regenerated by the transfer of a methyl group to homocysteine in a reaction catalyzed by methionine synthase (homocysteine methyl transferase) that requires methylcobalamin (B_{12}). Thus, low levels of B_{12} result in the accumulation of the substrate homocysteine.

Are there other conditions where homocysteine levels increase?

A deficiency of cystathionine β-synthase causes homocysteine to accumulate, but under these conditions, levels of methionine will also increase as a result of the increased remethylation of homocysteine via the methyl cycle (Figure 14.12).

Why does the elevation of homocysteine result in homocystinemia and homocystinuria?

Homocysteine readily oxidizes in the blood to homocystine (homocystinemia) and is excreted in the urine (homocystinuria).

How do the clinical manifestations of homocystinuria relate to the elevated levels of homocysteine in the serum?

The major clinical manifestations of patients with homocystinemia involve the eyes and the central nervous, skeletal, and vascular systems. It is believed that the clinical manifestations relate to impairment in collagen structure and function since homocysteine affects normal collagen crosslinking. Abnormal collagen crosslinking in this disease would account for the osteoporosis where the poor quality of ground substance for ossification leads to disrupted bone formation and increased turnover. Defective collagen formation also affects the eye, probably as a result of lax suspensory ligaments which permit lens dislocation. The propensity for significant thromboembolic events in these patients after the second decade of life most likely relates to the malformed intimal structure caused by abnormal collagen. In this latter case, patients have increased platelet turnover due to a continual process of adhesion to the abnormal intima. Overall, even mild hyperhomocystinemia is an independent risk factor for cardiovascular and atherosclerotic disease. Total plasma homocysteine values of approximately 10 μM for men and 8 μM for women are in the normal range. However, even a small increase of approximately 5 μM in total plasma homocysteine is associated with a 60% risk of coronary artery disease for men and 80% for women.

What is the molecular basis for pyridoxine (B_6) responsiveness in patients suffering from homocystinuria caused by a deficiency of cystathionine β-synthase?

Although the mechanism is not completely clear, it is thought that binding of the mutant cystathionine β-synthase to its cofactor (pyridoxal phosphate) may increase its stability, thus decreasing turnover within the cell and achieving higher absolute quantities of partially active enzyme. Evidence has shown that in responsive patients as much as 1,000 mg per day of pyridoxine may ameliorate many of the clinical consequences of the enzyme defect. Patients who are responsive achieve partial or complete normalization of plasma and urine methionine and homocysteine levels.

How else might one control homocystine levels?

In B_6-nonresponsive patients suffering from a deficiency in cystathionine β-synthase, dietary control of methionine intake is very important. In addition, controlled studies using betaine or choline to enhance remethylation of homocysteine to methionine have reduced homocystine levels in this group of patients.

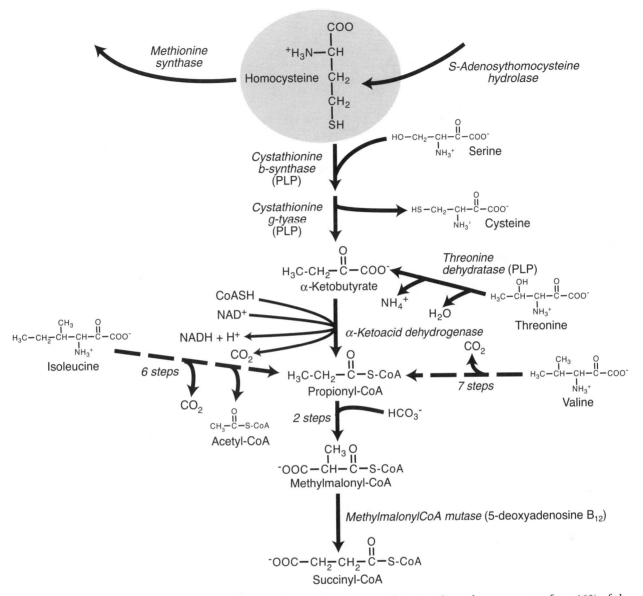

Figure 14.12 Succinyl-CoA pathway. Under normal conditions, cystathionine-β-synthase accounts for ~46% of the utilization of homocysteine. There is a significant reduction of cystathionine-β-synthase in a low protein diet, resulting in the predominant shunting of homocysteine back to methionine. S-Adenosyl-L-methionine activates cystathionine-β-synthase, which contributes significantly to the regulation of the succinyl-CoA pathway.

How does betaine and choline enhance the remethylation of homocysteine to methionine?

Betaine-homocysteine S-methyltransferase (BHMT) catalyzes a methyl transfer from betaine to homocysteine, forming dimethylglycine and methionine, respectively. Betaine is an intermediate of choline oxidation, and the enzymes of this pathway are primarily found in the liver and kidney of mammals. Although the significance of BHMT in homocysteine homeostasis is not clear, it is known that the relative contribution of BHMT to homocysteine remethylation can be influenced by diet. In fact, the rate of homocysteine remethylation is increased when choline or betaine is added to the diet of humans and will lower homocystine in non–B_6-responsive forms of homocystinuria (Figure 14.13).

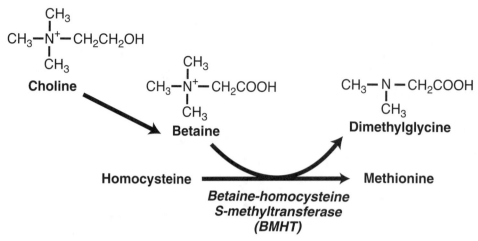

Figure 14.13 Betaine-homocysteine S-methyltransferase reaction. Betaine-homocysteine S-methyltransferase catalyzes a second alternative pathway, converting homocysteine to methionine in the liver. At essentially normal concentrations of substrates, betaine homocysteine methyltransferase accounts for ~27% of the utilization of homocysteine. The adaptation from a high protein diet to a low protein diet is achieved, in part, by a significant increase in betaine homocysteine methyltransferase to accommodate methionine demands.

Polyamine Metabolism

What is the essential role of polyamines in cellular function?

Polyamines such as spermine and spermidine are used in the packaging of DNA.

From what amino acids are polyamines derived?

Polyamines are synthesized from methionine and ornithine. The first step in their synthesis is the decarboxylation of ornithine (derived from arginine) catalyzed by ornithine decarboxylase, a pyridoxal phosphate-requiring enzyme.

What significance would inhibitors of polyamines have in the development of antiproliferative drugs?

Polyamines, spermine, and spermidine are used in DNA packaging, and they are required in large amounts in rapidly dividing cells; thus, inhibition of this enzyme would compromise a cells ability to proliferate. As an example, an inhibitor, difluoromethylornithine, which is relatively inert in solution, will bind ornithine decarboxylase and quickly inactivate the enzyme. This type of drug shows great promise in treating a wide range of diseases, particularly cancer-related diseases.

Phenylalanine, Tyrosine, and Tryptophan Metabolism

Why is the catabolism of phenylalanine related to tyrosine catabolism, and clinically important?

The breakdown of phenylalanine to tyrosine and ultimately to acetoacetyl-CoA and fumarate, which can enter the TCA cycle, is noteworthy because genetic defects in each of the first four enzymes in this catabolic pathway are known to cause inheritable human diseases.

When does tyrosine become an essential amino acid?

A deficiency in phenylalanine hydroxylase prevents the conversion of phenylalanine to tyrosine; thus, phenylalanine in the diet can no longer spare the requirement for tyrosine (Figure 14.14).

Why does a deficiency in phenylalanine hydroxylase lead to phenylketonuria?

When phenylalanine hydroxylase is defective, a secondary pathway of phenylalanine metabolism, normally little used, comes into play. In this minor pathway, phenylalanine undergoes transamination with pyruvate to yield phenylpyruvate. Both phenylalanine and phenylpyruvate accumulate in the blood and tissues and are excreted in the urine, hence the name of the condition phenylketonuria (PKU). Much of the phenylpyruvate is either decarboxylated to produce phenylacetate or reduced to form phenyllactate. It is phenylacetate that imparts the characteristic odor to the urine that has been used to detect PKU in infants (Figure 14.15).

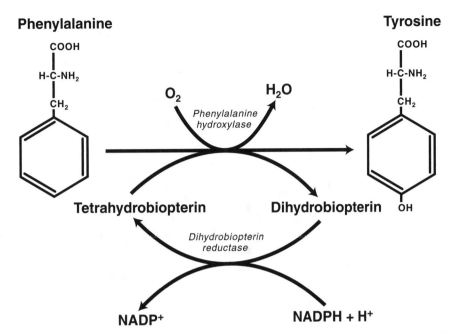

Figure 14.14 Phenylalanine hydroxylase reaction. Phenylalanine hydroxylase is dependent on the cofactor tetrahydrobiopterin. A deficiency in phenylalanine hydroxylase, which leads to phenylketonuria (PKU), is the most common inborn error of amino acid metabolism. Accumulation of phenylalanine or its metabolites in early life impairs the normal development of the brain, causing severe mental retardation. Excess phenylalanine may compete with other amino acids for transport across the blood brain barrier, resulting in a depletion of some required metabolites.

If not managed appropriately, what would be the outcome of a PKU infant?

The accumulation of phenylalanine or its metabolites in early life impairs the normal development of the brain, causing severe mental retardation. Excess of phenylalanine may compete with other amino acids for transport across the "blood–brain barrier," resulting in a depletion of some required metabolites.

How is PKU treated?

Treatment was indicated from the understanding of phenylalanine and tyrosine metabolism. When this condition is recognized early enough in infancy, mental retardation can largely be prevented by rigid dietary control. The diet must supply enough phenylalanine and tyrosine to meet the needs for protein synthesis. Consumption of foods that are rich in protein must be curtailed, and warnings are listed on foods artificially sweetened with "aspartame" (a dipeptide of the methyl ester of phenylalanine and aspartate). There is a national screening of newborns that has been highly effective, with a relatively inexpensive test and the detection and early treatment of PKU in infants (8 to 10 per 100,000).

What is the molecular basis for atypical PKU?

PKU can also be caused by a defect in the enzyme that catalyzes the regeneration of tetrahydrobiopterin called dihydrobiopterin reductase.

How does the treatment of atypical PKU differ from PKU?

In addition to the same rigid dietary control that was described for PKU, the atypical form would also require supplementation with L-dopa, 5-hydroxytryptophan, carbidopa, and folate to alleviate the patient's neurologic symptoms. Supplying tetrahydrobiopterin in the diet is insufficient because it is unstable and does not cross the blood–brain barrier.

Why is it necessary to supplement with L-dopa (levodopa), carbidopa, and 5-hydroxytryptophan in the treatment of atypical PKU?

The conversion of tyrosine to dopamine requires tetrahydrobiopterin, as well as the conversion of the tryptophan to 5-hydroxytryptophan. L-dopa is the immediate precursor to dopamine and can cross the blood–brain barrier in contrast to dopamine, which is unable to cross. When given alone, however, L-dopa is rapidly decarboxylated to dopamine in the gastrointestinal tract as well as in systemic circulation. In order to increase the levels of L-dopa effectively, carbidopa, which inhibits the systemic decarboxylase by binding to the pyridoxal binding site on the en-

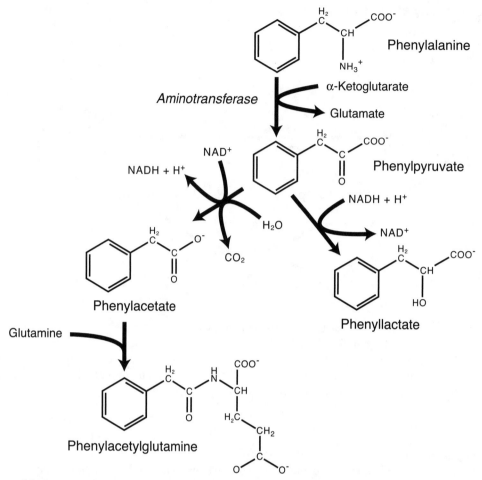

Figure 14.15 Phenylalanine metabolism in PKU. Deficiency of the enzyme phenylalanine hydroxylase leads to accumulation of phenylalanine in the plasma (>1200 mmol/L; reference range, 35–90 mmol/L) and to excretion of phenylpyruvic acid (approximately 1 g/day) and phenylacetic acid in the urine.

zyme and which is unable to cross the blood–brain barrier, is administered with L-dopa (levodopa) to enhance the L-dopa levels.

What are catecholamines?

The hormones epinephrine, norepinephrine, and dopamine are required for adaptation to acute and chronic stress.

Where are the catecholamines produced?

These amines are synthesized in the chromaffin cells of the adrenal medulla. Collections of these cells are also found in the heart, liver, kidney, gonads, adrenergic neurons of the postganglionic sympathetic system, and central nervous system. Catecholamines cannot cross the blood–brain barrier; thus, catecholamines of the brain are synthesized locally.

How is the production of catecholamine hormones regulated?

The rate-limiting enzyme in catecholamine biosynthesis is tyrosine hydroxylase. The activity of this enzyme is regulated primarily by feedback inhibition by the catecholamines (Figure 14.16).

Where do catecholamines act?

The catecholamines act through two major classes of receptors, designated α-adrenergic and β-adrenergic receptors. Each consists of two subclasses: α_1, α_2 and β_1, β_2. The affect on a tissue will depend on the relative affinity of the receptors. Epinephrine can bind and activate both types of receptors, whereas epinephrine binds primarily to the α-adrenergic receptors.

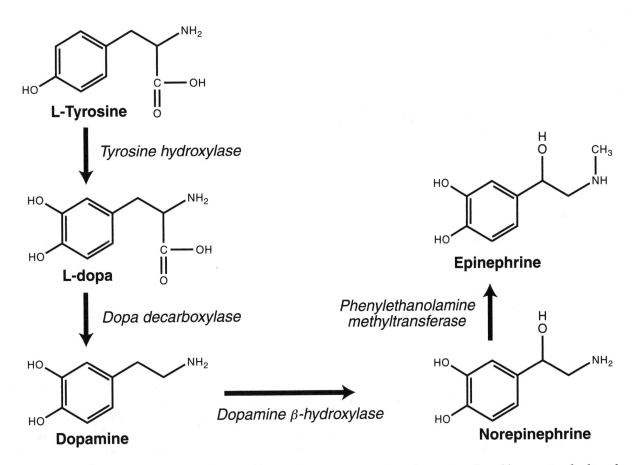

Figure 14.16   Catecholamine biosynthesis. The first step in catecholamine synthesis is catalyzed by tyrosine hydroxylase, which like phenylalanine hydroxylase, is dependent on tetrahydrobiopterin .Tyrosine hydroxylase produces DOPA. Subsequently, DOPA decarboxylase, with pyridoxal phosphate as a cofactor, produces dopamine, an active neurotransmitter, from DOPA. In the substantia nigra and some other parts of the brain, DOPA decarboxylase is the last enzyme in this pathway. In contrast, the adrenal medulla further converts dopamine to norepinephrine and epinephrine (adrenaline).

How are catecholamines inactivated?

Catecholamines are rapidly metabolized by catechol-O-methyltransferase and monoamine oxidase to form inactive compounds. Catechol-O-methyltransferase is found in many tissues and catalyzes the addition of a methyl group at the 3-position on the benzene ring. Monoamine oxidase deaminates monoamines and is also located in many tissues. However, the highest concentrations of monoamine oxidase are found in the liver, stomach, kidney, and intestine.

15 | Heme Metabolism

Where is the major site for heme biosynthesis?

The liver is the main nonerythrocyte source of heme synthesis.

Where is the heme molecule found?

Heme is a constituent of hemoglobin, myoglobin, and cytochromes.

What is the chemical nature of heme?

Heme is a porphyrin (a cyclic compound that contains four pyrrole rings linked together by methenyl bridges) (Figure 15.1).

How is heme synthesized?

The starting molecules from which heme is synthesized are glycine and succinyl-CoA. The rate-limiting enzyme in heme biosynthesis, δ-aminolevulinic acid synthase, is located in mitochondria and catalyzes the condensation of glycine and succinyl-CoA to form δ-aminolevulinic acid. Subsequently, in the cytosol, two molecules of δ-aminolevulinic condense to form a molecule containing a pyrrole ring called porphobilinogen. Four porphobilinogen molecules combine to form a linear tetrapyrrole, which cyclizes to uroporphyrinogen III and then coproporphyrinogen III. At this point, the pathway moves back into the mitochondrion where a series of carboxylation and oxidation reactions involving side chains in uroporphyrinogen III yield protoporphyrin IX. Iron is added to protoporphyrin IX by ferrochelatase to form heme (Figure 15.2).

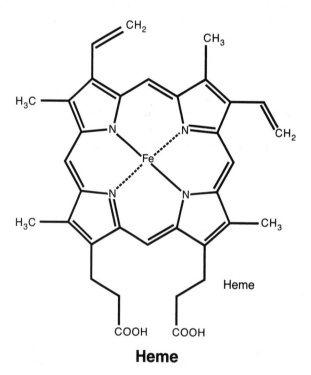

Heme

Figure 15.1 The structure of heme. Heme is a complex of protoporphyrin IX, a conjugated tetrapyrrole ring, and ferrous iron (Fe^{2+}).

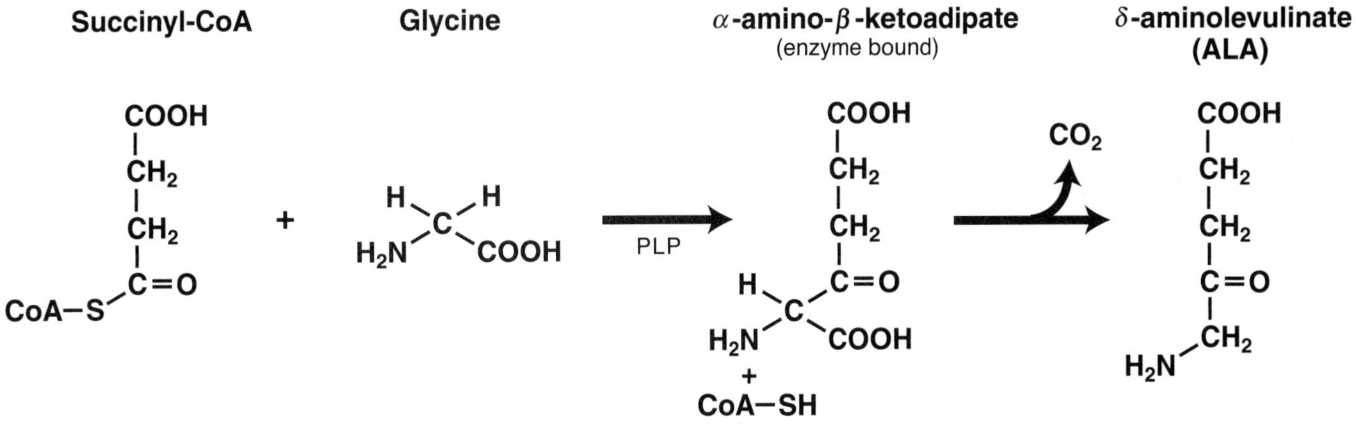

Figure 15.2 δ-Aminolevulinic acid synthase catalyzed reaction. δ-Aminolevulinic acid synthase catalyzes the conjugation of glycine with succinyl-CoA to yield α-amino-β-ketoadipate, which then decarboxylates to δ-aminolevulinate. Porphyrin biosynthesis is regulated by the concentration of heme, which feeds back and inhibits δ-aminolevulinic acid synthase.

How is heme biosynthesis regulated?

The product of the pathways, heme, feeds back and inhibits 5-aminolevulinic acid synthase.

What are porphyrias?

Porphyrias result from inborn errors that produce defects in the heme biosynthetic pathway. Different forms of porphyrias result from defects in enzymes starting with δ-aminolevulinic acid synthase and ending with ferrochelatase. Laboratory diagnosis of porphyrias involves both the measurements of metabolites of heme synthetic pathway and of relevant enzymes.

How is heme degraded?

The ring structure of heme is oxidatively cleaved to biliverdin by heme oxygenase, a P_{450} cytochrome. Biliverdin is enzymatically reduced to produce bilirubin. About 75% of bilirubin results from the turnover of red blood cells, which are phagocytosed by mononuclear cells of the spleen, bone marrow, and liver.

What is jaundice?

The normal plasma concentration of bilirubin is less than 17 μmol/L (1.0 mg/dL). Increased concentrations (more than 50 μmol/L or 3 mg/dL) impart a yellow color to the skin known as jaundice. Abnormalities in bilirubin metabolism may lead to elevation and are diagnostic of liver disease.

Why is jaundice diagnostic of liver disease?

Bilirubin is not very water soluble and must be carried in the blood complexed to albumin. When taken up by the liver, bilirubin is converted to a more soluble compound by esterification of one or both of its carboxylic acid side chains with glucuronic acid, xylose, or ribose. The glucuronide diester is the major conjugate, and its formation is catalyzed by a UDP glucuronyl transferase. Conjugated bilirubin is then secreted by the hepatocyte into the biliary canaliculi. Failure of a damaged liver to excrete conjugated bilirubin will therefore cause jaundice.

What happens to the conjugated bilirubin in the gut?

Bacteria in the gut further metabolize conjugated bilirubin to form stercobilinogen, also known as fecal urobilinogen, which is colorless. Stercobilinogen is oxidized to stercobilin (fecal urobilin), which is colored. Most all of the stercobilin is excreted in the feces and is responsible for fecal color.

Why are neonates susceptible to jaundice?

This is a transient condition that results from an accelerated hemolysis and an immature hepatic system for the uptake, conjugation, and secretion of bilirubin. Not only is the UDP-glucuronyl transferase activity reduced, but there is also frequently reduced synthesis of UDP-glucuronic acid. Exposure to visible light (phototherapy) can promote conversion of some bilirubin to other derivatives that are excreted in the bile.

16 | Nucleotide Metabolism

In addition to their role as a precursor for polynucleotide biosynthesis, what other important roles do nucleotides play in cells?

Nucleotides are important as energy carriers (i.e., ATP and GTP), cofactors (i.e., NAD and FAD), activated compounds (i.e., UDP-glucose and CDP-diacylglycerol), and second messengers (i.e., cAMP and cGMP).

What is the principal form of purine and pyrimidine compounds found in cells?

The principal form of purine and pyrimidine compounds found in cells is the 5'-nucleotide derivative. In normal cells, the nucleotide of highest concentration is ATP. Depending on the cell type, the concentrations of the nucleotides vary greatly. For example, in red blood cells, the adenine nucleotides far exceed the other nucleotides, which are barely detectable. In liver cells and other tissues, a complete spectrum of the monophosphates, diphosphates, and triphosphates are typically found.

Are ribonucleotide and deoxyribonucleotide concentrations in the cell equivalent?

The ribonucleotide concentration in a cell is in the millimolar (mM) range, whereas the concentration of deoxyribonucleotides in the cell is in the micromolar range (μM). The deoxyribonucleotides, however, are subject to major changes during cell growth, as compared with the ribonucleotides, which remain relatively constant.

Do all components of the ribonucleotides remain constant in the cell?

The concentration of ribonucleotides is fixed within rather narrow limits, although the concentration of the individual components can vary. For example, the total concentration of adenine nucleotides (AMP, ADP, and ATP) is relatively constant, although there may be variation in the ratio of ATP to AMP + ADP, depending on the energy state of the cell.

How does a cell maintain relatively constant levels of the ribonucleotides?

There is a finely regulated metabolic system that maintains the appropriate balance of all nucleotides. *De novo* synthesis of nucleotides begins with their metabolic precursors: amino acids, ribose-5-phosphate, CO_2, and NH_3. Salvage pathways recycle the free bases and nucleosides released from nucleic acid breakdown.

Purine Nucleotide Biosynthesis and Salvage

What is the committed step in the biosynthesis of purines?

The first committed step of the pathway is catalyzed by the enzyme glutamine PRPP amidotransferase. An amino group donated by glutamine is attached at C-1 of PRPP (phosphoribosyl-1-pyrophosphate). On the resulting 5-phosphoribosylamine, the purine ring is subsequently built (Figure 16.1).

How is glutamine PRPP amidotransferase regulated?

Glutamine PRPP amidotransferase may exist in one of two forms with molecular weights of 133,000 and 270,000. Enzyme activity is correlated with the smaller form. In the presence of 5'-purine nucleotides (i.e., 5'-AMP, 5'-GMP, or 5'-IMP), the smaller active form is converted to the larger inactivate form. The enzyme appears to have at least two effector sites. One binds AMP, and the other binds GMP or IMP. The simultaneous binding of both results in a synergistic inhibition of the enzyme. In the presence of PRPP, the large form is shifted to the more active smaller form.

What is the source of PRPP?

PRPP is synthesized by 5-phosphoribose-1-pyrophosphate kinase (PRPP synthetase) using α-ribose-5-phosphate and ATP. The reaction requires Mg^{2+} (Figure 16.2).

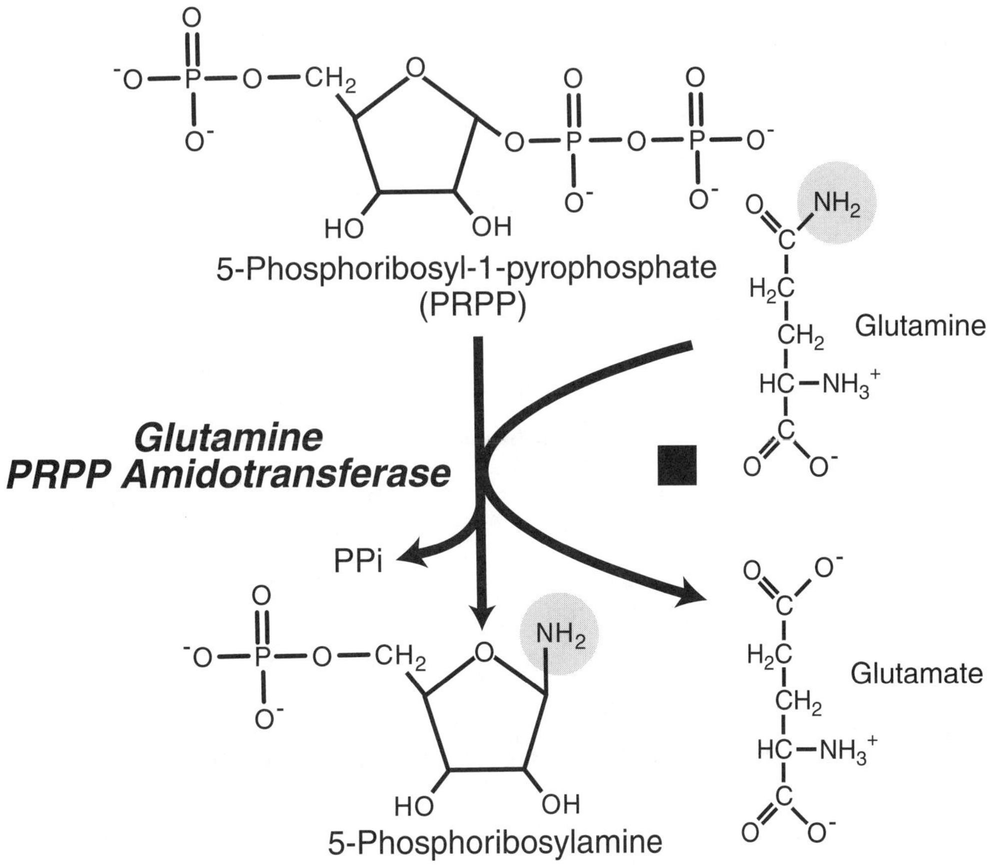

Figure 16.1 Glutamine PRPP amidotransferase reaction. The first committed step in *de novo* purine biosynthesis is catalyzed by glutamine PRPP amidotransferase where glutamine donates an amino group to the C-1 position of PRPP to produce 5-phosphoribosylamine.

Where does the ribose-5-phosphate come from?

The ribose-5-phosphate is generated from glucose-6-phosphate via the hexose monophosphate shunt or from ribose-1-phosphate (generated by phosphorolysis of nucleotides) via a phosphoribomutase reaction.

Is the synthesis of PRPP regulated?

The activity of PRPP synthetase is inhibited by nucleoside diphosphates and triphosphates. ADP is the most potent inhibitor of PRPP. ADP is a competitive inhibitor to ATP at concentrations equivalent to intracellular concentrations.

What is the metabolic significance of IMP?

IMP serves as the common precursor for GMP and AMP synthesis. The synthesis of AMP and GMP from IMP is based on cellular needs through feedback regulation at the IMP branch point.

How is this balance for GMP and AMP biosynthesis maintained?

Regulation occurs at the IMP branch point. Although the two enzymes, IMP dehydrogenase and adenylosuccinate synthetase, that use IMP at this branch point have a similar K_m for IMP, AMP feeds back to inhibit adenylosuccinate synthetase competitively, and GMP feeds back to inhibit IMP dehydrogenase competitively (Figure 16.3).

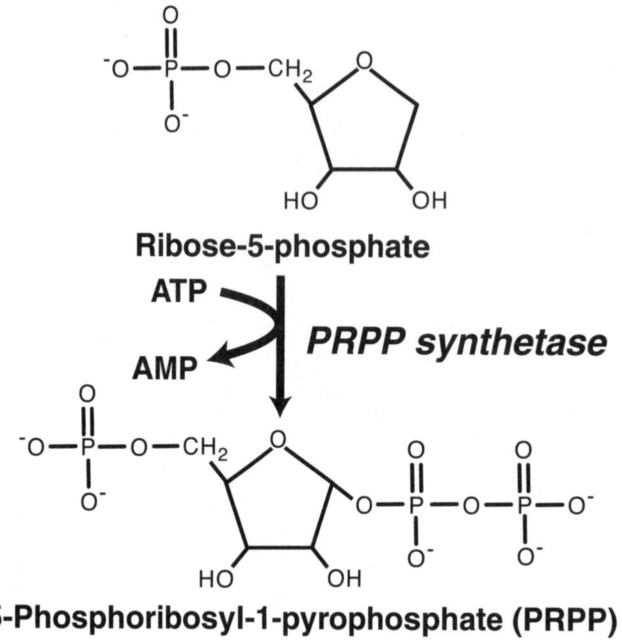

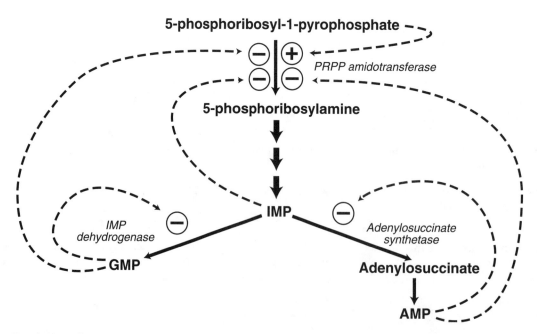

Figure 16.2 PRPP synthetase reaction. PRPP is synthesized from ribose-5-phosphate derived from the pentose pathway in a reaction catalyzed by ribose PRPP synthetase. This enzyme is allosterically regulated by biomolecules for which PRPP is a precursor.

Figure 16.3 Regulation of purine biosynthesis. There are two major sites of feedback regulation that affect the overall rate of *de novo* purine nucleotide synthesis. The first is at the committed step catalyzed by glutamine PRPP amidotransferase, and the second site is at IMP conversion to either GMP, catalyzed by IMP dehydrogenase, or AMP, catalyzed by adenylosuccinate synthetase.

What is the metabolic significance of the ability of cells to carry out the salvage of purines?

De novo synthesis of nucleotides requires a relatively high energy requirement; therefore, most cells make use of very efficient salvage pathways that allow purine and pyrimidine bases to be reused. To synthesize a purine nucleotide requires a minimum of six ATPs; thus, to salvage the purine ring is more economical. Therefore, *de novo* synthesis plays a more important role during tissue growth and development under conditions in which there is a high demand for nucleotides.

Because of the complexity of the pathway for purine biosynthesis, why are inborn errors in *de novo* biosynthesis of purines uncommon?

This may simply relate to the fact that an inability to carry out *de novo* synthesis is incompatible with life.

What enzymatic mechanisms catalyze the salvage of purines?

There are two distinct enzymes involved in salvage of the purine base. Hypoxanthine-guanine phosphoribosyltransferase (HGRPTase) uses PRPP to salvage the hypoxanthine and guanine purines. Adenine phosphoribosyltransferase (APRTase) salvages adenine.

How does the salvage of the purine bases regulate *de novo* biosynthesis?

The generation of AMP and GMP through their respective salvage pathways effectively shuts off *de novo* synthesis at PRPP amidotransferase in two ways. First, salvage uses PRPP, thus decreasing PRPP available for PRPP amidotransferase. Second, the generation of AMP and GMP via their salvage pathway serves to feed back and inhibit glutamine PRPP amidotransferase.

Why is the salvage of purine bases critical to the red blood cell?

Red blood cells lack glutamine PRPP amidotransferase and therefore cannot carry out *de novo* purine synthesis. Therefore, the red blood cell is dependent on salvage pathways to maintain purine nucleotide pools.

What other means do cells have to balance the relative distribution of adenine and guanine nucleotides?

Purine nucleotides can be redistributed to meet cellular needs through the conversion of GMP and AMP back to IMP. GMP reductase catalyzes the conversion of GMP to XMP, and this conversion is regulated because it is inhibited by XMP and activated by GTP. AMP deaminase converts AMP to XMP, and this conversion is regulated because it is inhibited by GTP and GDP, and activated by ATP. Interestingly, these two enzymes, GMP reductase and AMP deaminase, are usually associated with purine catabolism, yet they also serve a role in the interconversion and balance of purine nucleotides.

Purine Degradation

Which organ plays a primary role under normal conditions in the degradation of purines?

Degradation of purines occurs primarily in the liver.

Is there a common product derived from the degradation of purines?

Purine nucleotides, nucleosides, and bases share a common degradative pathway. Purine nucleotides are broken down to hypoxanthine and xanthine, which in turn are oxidized to uric acid by xanthine oxidase (xanthine dehydrogenase). Uric acid is the end-product of purine catabolism in humans (Figure 16.4).

How do adenosine deaminase and AMP deaminase differ?

AMP deaminase is much more substrate specific than adenosine deaminase, where the latter will recognize deoxyadenosine and many other 6-aminopurine derivatives. The significance of this difference in specificity means that the degradation of deoxyadenosine and its nucleotides is totally dependent on adenosine deaminase.

Why does a deficiency in adenosine deaminase lead to an immunodeficiency disease?

Adenosine deaminase deficiency (ADA), an autosomal recessive disease, seems selectively toxic to lymphocytes, in part based on their dependence on the salvage pathways. Patients with ADA typically lack both cell-mediated (T cell and humoral B cell) immunity, presenting with severe combined immunodeficiency disease. In the plasma and urine of ADA-deficient children, substrates of adenosine deaminase accumulate, in particular deoxyadenosine. There are several proposed mechanisms for the pathogenesis of the immune defect in ADA deficiency. Two of several proposed mechanisms for the molecular basis for pathogenesis are based on the accumulation of deoxyadenosine and its nucleotide derivatives. For example, the preferential accumulation of dATP in ADA-deficient T-lymphocytes has been suggested to inhibit ribonucleotide reductase, thus preventing deoxyribonucleotide synthesis, which would lead to inhibition of DNA replication. The accumulation of deoxyadenosine would also lead to the inhibition of S-adenosylhomocysteine

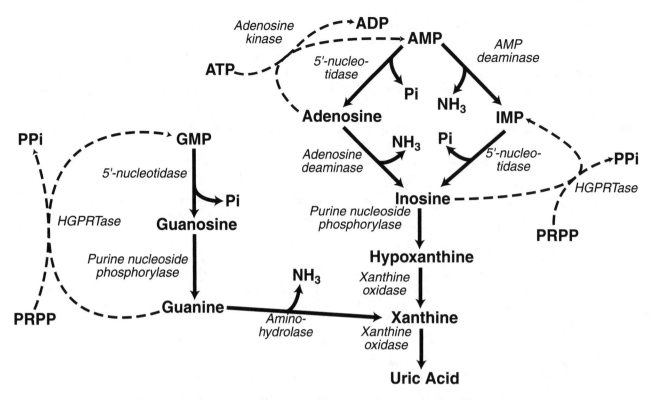

Figure 16.4 Purine catabolism pathway. Purine nucleotides are degraded by a pathway in which they lose their phosphate through the action of 5'-nucleotidase, and then their sugar group by the action of purine nucleoside phosphorylase. The loss of amino groups from the purine rings are catalyzed by either adenosine deaminase for adenosine, or AMP deaminase for AMP. In the case of guanine, deamination is catalyzed by aminohydrolase. Xanthine oxidase catalyzes the final steps of the pathway, converting hypoxanthine and xanthine to uric acid.

hydrolase and accumulation of S-adenosylhomocysteine. As a consequence, high levels of S-adenosylhomocysteine would lead to significant inhibition of all transmethylation reactions using SAM as the methyl donor, which in turn would significantly impair vital functions in affected lymphocytes.

Does purine nucleoside phosphorylase participate both in the degradation of purines as well as their salvage?
Although the equilibrium for the reaction catalyzed by nucleoside phosphorylase favors nucleoside synthesis, its major role in tissue appears to be in purine catabolism. This role is favored because the concentration of metabolites likely drives the reaction in the degradative direction.

What is the difference between xanthine oxidase and xanthine dehydrogenase?
Xanthine dehydrogenase catalyzes the conversion of hypoxanthine to xanthine and xanthine to uric acid:

$$\text{Hypoxanthine} + \text{NAD}^+ + \text{H}_2\text{O} \rightarrow \text{Xanthine} + \text{NADH} + \text{H}^+$$
$$\text{Xanthine} + \text{NAD}^+ + \text{H2O} \rightarrow \text{Urate}^- + \text{NADH} + 2\text{H}^+$$

Xanthine dehydrogenase can be converted to xanthine oxidase through the oxidation of cysteine thiols. Xanthine oxidase catalyzes the following reactions:

$$\text{Hypoxanthine} + \text{O}_2 + \text{H}_2\text{O} \rightarrow \text{Xanthine} + \text{H}_2\text{O}_2$$
$$\text{Xanthine} + \text{O}_2 + \text{H}_2\text{O} \rightarrow \text{Urate}^- + \text{H}^+ + \text{H}_2\text{O}_2$$

The xanthine dehydrogenase and xanthine oxidase reactions, catalyzed by the same protein, differ, but many texts use the terms interchangeably. The xanthine oxidase reaction results in the production of hydrogen peroxide, and the xanthine dehydrogenase reaction does not. The reaction catalyzed by xanthine dehydrogenase is the physiologically important process in purine degradation *in vivo*.

Is there any physiologic significance associated with the conversion of xanthine dehydrogenase to xanthine oxidase?

The conversion of xanthine dehydrogenase to xanthine oxidase occurs within hypoxic cells. This metabolic transformation produces xanthine oxidase that generates toxic oxygen intermediates such as superoxide and peroxides.

What common disease state is related to purine degradation?

A key physical property of uric acid is its limited solubility in serum. Gout is a clinical disorder characterized by elevated serum uric acid levels and deposits of sodium urate in the joints of the extremities.

What are the essential causes of gout?

Primary gout is a form that results from inborn error of metabolism and is biochemically and genetically heterogenous. For the most part, the largest subgroup consists of patients in whom the biochemical defect is undefined. Many factors can influence the presentation of gout in these patients, such as reduction in renal clearance, dietary factors (excesses in calories), and alcohol ingestion. Defects in PRPP synthetase and partial deficiency in HGRPTase are two examples of defects resulting in primary gout. Secondary gout results from an increase in *de novo* purine, which may be associated with glucose-6-phosphatase deficiency (Von Gierke's disease or type I glycogen storage disease) and complete HGRPTase deficiency (Lesch-Nyhan syndrome). In addition, secondary gout may also result from tissue damage or decreased excretion associated with renal dysfunction where there is a significant increase in nucleic acid turnover.

Why would a defect in PRPP synthesis lead to gout?

HGRPTase uses PRPP to salvage the hypoxanthine and guanine purines, and APRTase salvages adenine. Without sufficient PRPP the purine base will be degraded to form uric acid.

Why would a defect in PRPP use lead to hyperuricemia and gout?

Because more than 90% of purine bases are salvaged for recycling, a decreased HGRPTase would result in an increase of purine catabolism and uric acid production. Additionally, there would be an elevation of PRPP because it would no longer be used by a deficient HGRPTase. The increase in PRPP would, in turn, promote *de novo* purine biosynthesis because a significant elevation of PRPP would overcome the feedback regulation of PRPP amidotransferase by AMP, GMP, and IMP. Finally, because guanine and hypoxanthine are not salvaged, the nucleotide products cannot feed back and inhibit *de novo* purine synthesis.

Why does allopurinol, an inhibitor of xanthine oxidase, have little affect on the abnormalities of Lesch-Nyhan syndrome, other than to reduce uric acid levels?

In Lesch-Nyhan syndrome, PRPP levels increase dramatically stimulating *de novo* synthesis, while feedback inhibition by IMP and GMP on *de novo* synthesis as a result of salvage is lost. Moreover, allopurinol itself relies on HGRPTase for salvage in order to act as a purine nucleotide analogue capable of directly inhibiting PRPP amidotransferase, and therefore *de novo* purine biosynthesis.

Pyrimidine Biosynthesis

How does the strategy for *de novo* pyrimidine ring synthesis in mammalian cells differ from purine *de novo* biosynthesis?

The pyrimidine ring is not assembled on the sugar moiety, but rather independently. After the pyrimidine ring is formed, it is conjugated with the ribose ring via PRPP.

What features of the pyrimidine ring biosynthesis are common with purine ring biosynthesis?

The pyrimidine ring is synthesized *de novo* in mammalian cells using amino acids as carbon and nitrogen donors and CO_2 as a carbon donor (Figure 16.5).

What is the significance of multifunctional enzymes in the *de novo* synthesis of pyrimidines?

Although the *de novo* pathway for UMP synthesis requires six enzyme activities, these activities reside on only three gene products. Carbamoyl phosphate synthetase, aspartate carbamoyl transferase, and dihydroorotase activities are present on the same polypeptide chain located in the cytosol of the cell. Dihydroorotate dehydrogenase activity is on a separate protein that resides in the mitochondria, and orotate phosphoribosyl transferase and OMP-decarboxylase are on the same polypeptide found in the cytosol, referred to as UMP synthase. As a result of the channeling of intermediates through these enzyme steps, essentially none of the metabolites between the first step and the last step is found in the intracellular pool of the cell.

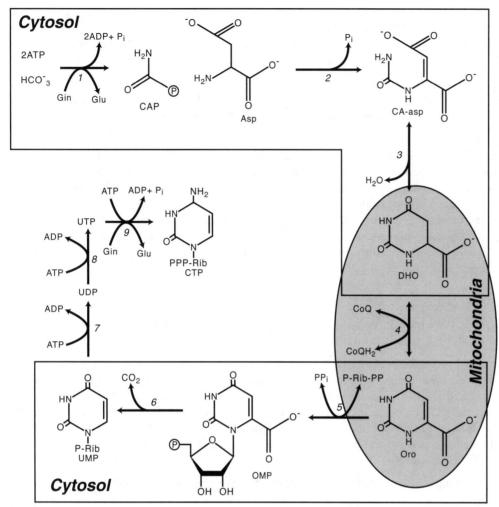

Cytosol

1. Carbamoyl-phosphate synthase II

2. Aspartate trans-carbamoylase

3. Dihydroorotase

4. Dihydroorotase dehydrogenase

5. Orotate phosphoribosyl transferase

6. Orotic acid decarboxylase

7 & 8. Kinases

9 CTP synthetase

NOTE:
Enzyme activities 1, 2, 3 represent a multifunctional enzyme.

Enzyme activities 5 & 6 represent a multifunctional enzyme called *UMP Synthase.*

Figure 16.5   Pyrimidine nucleotide biosynthesis. Pyrimidines are constructed from carbamoyl phosphate and aspartate. In humans, the regulated step in this pathway is catalyzed by carbamoyl phosphate synthetase II.

How is the product, UMP, from *de novo* pyrimidine biosynthesis converted to the cytidine nucleotide?
   Nucleotide kinase converts UMP to UTP, which is then used by the enzyme CTP synthetase to produce CTP.

CTP synthetase
UTP + Glutamine + ATP → CTP + Glutamate + ADP + Pi

Where does the regulation of pyrimidine biosynthesis occur in mammalian cells?
   Unlike bacterial cells where regulation occurs at the aspartate carbamoyl transferase step, the regulation of mammalian pyrimidine nucleotide synthesis occurs at the level of carbamoyl phosphate synthetase II, which is inhibited by UTP. The next step of regulation is UMP synthase, where UMP, and to a lesser extent, CMP are inhibitors of the OMP-decarboxylase activity. In addition, CTP down regulates CTP synthase; thus, the extent of conversion of UTP to CTP is based on relative cellular concentrations (Figure 16.6).

Because OMP-decarboxylase and orotate phosphoribosyl transferase reside on the same polypeptide chain of the enzyme commonly referred to as UMP synthase, what is the significance of the feedback inhibition by UMP at this enzyme?
   Under physiologic conditions where there is excessive ammonia, the liver, through carbamoyl phosphate synthetase I, detoxifies NH$_3$ by forming carbamoyl phosphate in the mitochondria. Under pathologic conditions, excessive carbamoyl phosphate passes into cytosol and becomes a substrate for aspartate carbamoyl transferase in pyrimidine *de novo* biosynthesis pathway. Importantly, this source of carbamoyl phosphate bypasses the reaction catalyzed by carbamoyl phosphate synthetase II, the regulated step in pyrimidine biosynthesis. Because of channeling, this ultimately results in the increased production of orotic acid because UMP will feedback and inhibit UMP synthase, thus pre-

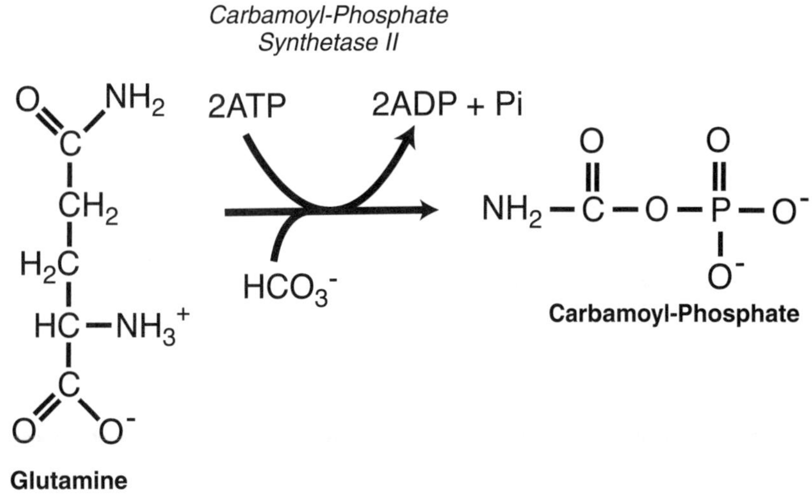

Figure 16.6 Carbamoyl phosphate synthetase II catalyzed reaction. Carbamoyl phosphate synthetase II is located in the cytosol and utilizes glutamine as the amino donor.

venting conversion of orotic acid to UMP. Thus, patients with ammonia toxicity present with high levels of orotic acid in their urine (orotic aciduria).

What features distinguish carbamoyl phosphate synthetase II involved in pyrimidine biosynthesis from carbamoyl phosphate synthetase I involved in urea biosynthesis?

Carbamoyl phosphate synthetase II (CPSII) is a cytosolic, multifunctional protein with three different catalytic sites mediating the three initial reactions of *de novo* pyrimidine biosynthesis. CPSII uses glutamine rather than ammonia as the donor of the amine groups of the carbamoyl phosphate and is inhibited by UTP and CTP. Also, CPSII does not require N-acetylglutamate as a positive allosteric regulator for activity.

Are there any inborn errors in metabolism associated with *de novo* pyrimidine biosynthesis?

Essentially, only one enzyme defect in pyrimidine biosynthesis has been recognized, involving an autosomal recessive deficiency of the bifunctional enzyme, UMP synthase. Because of the critical role of *de novo* pyrimidine biosynthesis in embryogenesis, there have been no recognized complete deficiencies in UMP synthase observed. UMP synthase deficiency presents clinically with orotic aciduria, but in this instance, there is no elevation of serum ammonia. In addition, there is the eventual development of megaloblastic anemia. The megaloblastic anemia results from impairment in the normal rate of cell division of the red blood cell precursors in bone marrow.

Can one differentiate orotic aciduria associated with a urea cycle defect (ornithine transcarbamoylase deficiency) from that associated with a UMP synthase deficiency?

Key differential points are the blood ammonia, which is normal in UMP synthase deficiency, and severe with a urea cycle defect.

Are there salvage pathways for the pyrimidine bases?

Pyrimidine bases can be salvaged by reactions involving pyrimidine phosphoribosyltransferase. Although the salvage of pyrimidines has not been well characterized, there is an enzyme activity identified from human erythrocytes that has been shown to use orotate, uracil, thymine, but not cytosine.

Pyrimidine phosphoribosyltransferase
Pyrimidine + PRPP→ Pyrimidine nucleoside monophosphate + PPi

Deoxynucleotide Formation

What is the key step in the formation of deoxyribonucleotides?

Deoxyribonucleotides are formed directly from the reduction of the 2'-position of the corresponding ribonucleotides by ribonucleotide reductase. This reduction is specific for nucleoside diphosphates, and the reaction is regulated so as to maintain the appropriate balance of deoxyribonucleotides (Figure 16.7).

How is the formation of deoxyribonucleotides regulated?

Ribonucleotide reductase is regulated in two ways: by the level of enzyme expressed and by allosteric regulation that balances the levels of the deoxyribonucleotide products. The fact that deoxy ATP is a potent inhibitor of the reduction of all four nucleoside diphosphate substrates also provides some explanation for the toxicity of deoxyadenosine build up in ADA.

Because dTDP is not a product of ribonucleotide reductase, how are the deoxythymidine nucleotides synthesized?

Deoxy-TMP (dTMP) is synthesized from dUMP by the enzyme thymidylate synthase (Figure 16.8).

What is the source of dUMP?

The major source of dUMP appears to result from deamination of dCMP, rather than from the conversion of dUDP to dUMP.

Why and how is the production of dUTP prevented?

Deoxy-UTPase, with a K_m for dUTP of approximately 1 μM, converts dUTP to dUMP, thus preventing the incorporation of dUTP into DNA during DNA replication or repair.

$$UTPase$$
$$dUTP \rightarrow dUMP + PPi$$

How are the pyrimidine bases catabolized?

CMP can be deaminated to UMP by CMP deaminase, or cytosine can be deaminated by a different enzyme, cytosine deaminase, to uridine. Subsequently, pyrimidine nucleosides (uridine or thymidine) are broken down by nucleoside phosphorylase to the free base (uracil or thymine), releasing ribose-1-phosphate.

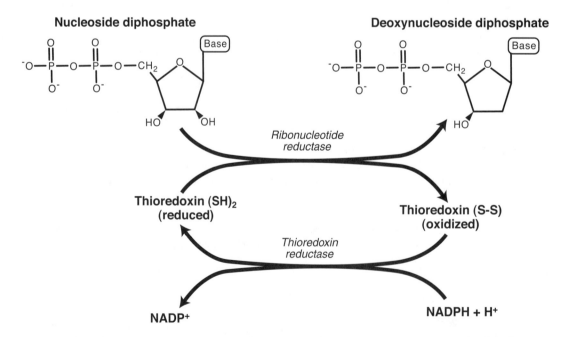

Figure 16.7 Ribonucleotide reductase reaction. The reduction of the D-ribose portion of a ribonucleoside diphosphate to 2'-deoxy-D-ribose is catalyzed by the enzyme ribonucleotide reductase. Thioredoxin serves as a reductant and has pairs of -SH groups that carry hydrogen atoms from NADPH.

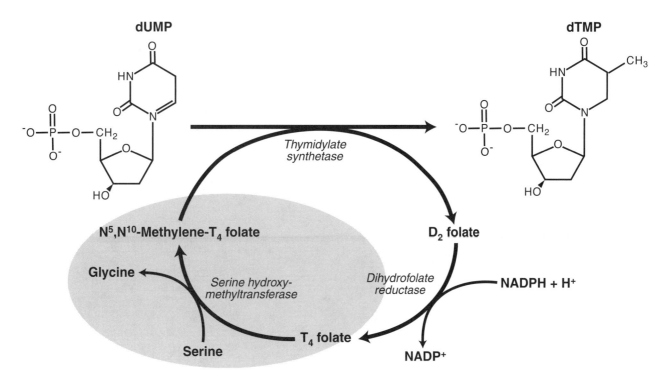

Figure 16.8 Thymidylate synthetase reaction. Conversion of dUMP to dTMP is catalyzed by thymidylate synthetase. The one-carbon unit is transferred from N^5,N^{10}-methylenetetrahydrofolate to dUMP.

Are the pyrimidine bases degraded by a common pathway?

Although uracil and thymine are degraded in a similar fashion, they yield different products. Uracil is degraded to β-alanine, NH_4^+, and CO_2, whereas thymine is degraded to β-aminoisobutyric acid, NH_4^+, and CO_2. Because thymine is degraded to a unique product, β-aminoisobutyric acid, which is excreted in the urine, this substance can be used to estimate DNA turnover (tissue damage).